HOTHOUSE TRANSPLANTS

Stories collected by Matt Duffy

Grove Publishing
Westminster, California

Grove Publishing
16172 Huxley Circle
Westminster, CA 92683
(714) 841-1220

ISBN 0-929320-12-3
Copyright 1997 Grove Publishing

Printed in the United States of America

INTRODUCTION

Many of you might be wondering where the title *Hot-House Transplants* came from—a very good question. The answer begins with the idea for the book itself. My mother and I attended the National Home School Leadership Conference in Orlando, Florida a few years ago and were sharing a room with Mrs. Marilyn Rockett and her son Jeremy. I believe it was the last evening of the conference at around 11:00 p.m. when we were talking about some of the great things that homeschool graduates were doing. One or more of us came up with the bright idea of putting together a book containing stories by homeschoolers, telling what they are doing and have done as a result of homeschooling. The end product of that conversation is what you now hold in your hands, *HotHouse Transplants*.

As for the title itself, homeschooling parents have often been criticized for sheltering their children, for raising them in a "hothouse" environment where they are not exposed to the "real world." To a certain extent this is true, although most homeschooling parents will present their case in two ways: my child actually gets to see more of the real world because of all the activities we participate in, and/or children really are like tender, baby plants that need protection until they are mature enough to confront all of the evil that is out there.

HOTHOUSE TRANSPLANTS

Since home-educated students are, in a way, transplanted from homeschooling into the "real world," the title seemed rather appropriate. It is my hope that this book will be encouraging to those of you who are homeschooling at present or to those of you who may be looking at homeschooling as a possibility.

Matt Duffy

ACKNOWLEDGEMENTS

First, I would like to thank all those who contributed stories to this book. They generously shared their stories, hoping it will encourage those who read them.

I wouldn't be doing what I'm doing if it weren't for many people who have influenced and helped me, so I would like to thank Kathleen Courtney, Valerie Thorpe, Mr. and Mrs. Craig Roberts, Diane Eastman, Dennis White, Joy Feria, Frank Tretter, and Kristi Tretter for all the classes and lessons through which they taught and inspired me.

An extra thank you goes to Diane Eastman and Katheleen Courtney for helping to edit this book. Chris Duffy and Rachel Thorpe get credit and thanks for the fantastic cover art and graphics.

A special thanks to my parents for all their guidance, wisdom, and nagging.

And most of all, I thank God for helping me to actually get this project completed.

Matt Duffy

Matt Duffy lives (at least some of the time) with his parents and two older brothers in Westminster, California. He graduated from homeschool in June, 1997.

4

CHAPTER 1

Call me a social creature.... But I never got to see the inside of a classroom until I was sixteen years old. I've never had the blessing of sitting through a class next to an irregularly dressed member of some unique social group. I have never enjoyed the gut-wrenching feelings most guys experience when asking a girl to a dance, to study with you, or out on a date for that matter (although skipping dating isn't an automatic part of home education). I've never had to worry about remembering my locker combination. And I've never had to wait in line to use the bathroom. I feel so deprived!

I made the mistake once, probably when I was seven or eight, of telling my parents that I wanted to go to a public school to "see what it was like." Little did I know the reaction I would get. My mother started laughing. My two older brothers (who had been in school a little bit) laughed too and, almost in unison, responded, "No, you don't!" I never asked again. Yet, my curiosity had been tweaked each time I talked with public school friends as they discussed the wonderful things they got to do in their classes. I remember an instance, several months ago, when one friend told me that his history teacher rented the movie *Braveheart* which they watched for their class for two days. Another explained to me how they took a trip to Disneyland so they could, as the teacher said, "Be exposed to many cultures."

Then it hit me. I realized why the public school system seemed so appealing to me. It was easier. Being exposed to nothing but the educational resources that my mother used for me and my brothers meant that I had experienced quite a bit of difficult work in challenging subjects. The education some of my friends were getting seemed so much more fun. It was inviting. I tell you all of this because I want you to understand what kind of education I have had.

I have homeschooled for. . . well, . . . forever. And I honestly would not trade it for anything. Through home education, I have had the opportunity to do so many things which I would not have been able to do in the public educational system or even in a traditional Christian school. For instance, I have, on occasion over the past eight or nine years, been a salesman for my mother's business. It is because of this that I was able to travel throughout the country, including Hawaii, and sell our homeschool products. It is also because of this that I have decided that I will probably pursue a Business Management degree in college.

Still, people wonder how you can be normal if you are home educated. I have been asked often by my peers if I wouldn't rather be at a "real" high school where I could have a social life, where I could go to the dances and the prom, where I could join the school athletic teams, and so forth.

The idea of having a social life is greatly misinterpreted by those not personally familiar with homeschooling. I simply have to laugh when people try to say that I don't have a social life. Everyone who knows me, would agree that I do not lack for social activity. My idea of a good week is to have events happening 15 to 20 hours a day, back to back; and, often, that is how some days are. For this reason I have been dubbed a "social creature" by my parents, and rightly so.

To replace the traditional sports others take part in through their schools, I regularly play my favorite sport, volleyball, with family and friends. I've played for two years in a church volleyball league. I participated in a P.E. program provided by the local YMCA with a group of other homeschoolers and in sports programs provided by my city. I have always been very involved with church, Boy Scouts, work, and so many other "social" activities, that my parents have to regularly restrict me from doing too much.

When it comes to school work, I would like to have had easier work than what I was given by my mother and other teachers, but would it have been of real benefit? Most definitely not. In the long run, I am finding that the advantages and overall blessing which have come from home education far outweigh the appeals of other educational options.

Now it's time for me to figure out what to do with the rest of my life. But before I can go on with my story, I have to tell you that for the first 15 to 16 years of my life, my friends and I would joke and call each other "politicians" to mock each other. It was our ultimate insult! In those days, all I knew of politics was that practically everyone involved was "evil." God wanted to change my thinking, and perhaps my life direction as He often does. My mother (Cathy Duffy) had the opportunity to meet with a gentleman who was running for our State Assembly

District seat. (He wanted to find out if homeschoolers would support him.) My mother was very positively impressed with him, and many of our local homeschoolers got involved with his campaign and helped him get elected.

He had formerly been a teacher, enjoyed working with young people, and mentioned the possibility of having some home educated teens come work in his office as interns. The first time my mother mentioned it, I wasn't interested or even really listening. (Actually, I don't remember the first time it came up; this is from my mother's recollection of the chain of events.)

Later that month, my mother ran into a young man at a Republican event who also worked with our assemblyman—the notion of interning came up again. Thank God for these "chance" meetings! By that time, I was starting to listen and seriously consider the possibility of actually getting involved in politics.

I met that same young man at the 1996 Republican celebration and discussed with him the possibility of interning for my assemblyman. I called their office and was brought into the ever crazy, changing, hectic, eventful, challenging, yet almost always fun world of politics. There would be no more insults about politicians from me; I am slowly becoming one.

Largely due to the people I am now working for and with in my state assemblyman's office, my love for politics has grown so much that I am determined to eventually run for the State Assembly seat for our district.

I realize that at the young age of 18, my life is far from over, and many things will still change. But, I find it interesting that politics, the last thing in the world I would have ever thought of becoming involved in a few years ago, has become practically a passion. My present job with my assemblyman is the first step toward my future.

I will be taking the second step along this path this summer. It is my privilege to have been accepted to be an intern at the Home School Legal Defense Association for the next six months. Though I have not yet gone, I already know that the experience of working with such dedicated people as Chris Klicka, Michael Smith, Michael Farris, Matt Chancey (whose story you can read in Chapter 5), Scott Sommerville, and numerous others is going to most assuredly be the time of my life. I have no doubt that anyone considering working in the political arena or becoming involved with government issues as a concerned citizen should get involved with HSLDA.

The Bible describes examples of many people who struggled with giving their whole life to God for His purposes. I am one of those who continually struggles. But I think God has chosen me for politics for several reasons. Our government is evil. Not all of it, but much of it has become a pool of greed, power, and wealth. There is a lack of morality in our country. Things God finds repulsive are being made into law without any qualms!

There are so many problems in our country, and if Christians would just get more involved, this country could be changed. The hearts of the people could be softened to morality. This is my calling.

This book is also about God's calling for many young Christians who have entered the world. Every story in this book stresses the wonderful advantages and opportunities of home education. The stories are about education, courage, and life choices; they are also about God.

There is one more person's story I wish were in this book to make it complete—an example of how home education has brought forth a man of knowledge and integrity. Michael New has stood up for the Constitution of the United States in the face of the most challenging powers of this world. In 1993, Specialist Michael New, a member of the United States Army was ordered to wear the United Nations insignia on his uniform and, thereby, become a United Nations soldier. Michael New understood his agreement with the United States, how he swore to protect it and uphold it. He stood against unconstitutional orders and faced the entire United States Army as well as the United Nations. Michael New was like every other man in that army, but for one thing—he was homeschooled. Through homeschooling he gained the knowledge and courage to stand for what is right. He has made an impact in our country the likes of which will be felt for years to come. He is an example of a true homeschool success story.

So read on. Whether you are already homeschooling or are considering it, be encouraged by the trials these fellow Christians and homeschoolers have faced. Be encouraged by their testimony of homeschooling. Be encouraged by the many ways God has worked through them.

And for any teens or young adults reading this book who are wondering what they are going to do with their lives—or even if God can use them—just remember, everyone has a place in the kingdom of God. You don't have to be a CEO, a president, or even a pastor. . . God uses everyone. Even if you don't know it yet, you are already God's success story if you have a willing and obedient heart.

HOTHOUSE TRANSPLANTS

Photos by Debbie Trueblood

Jonathan Rockett attends Hillsdale College in Hillsdale, Michigan. When he is on break from school, he lives with his family in Waldorf, Maryland.

CHAPTER 2

"Hmmm... A superior education." As I was getting ready for work one morning I contemplated my current job status. "Yeah. A superior education is the ingredient that has brought me to this high rung of the executive ladder." I glued my bulbous, red nose into place and headed out the front door. A voice from behind bade me farewell, "Good-bye, Bananas." I climbed into the car, backed out of the driveway, and headed down the road to a house where a dozen anxious children waited for Bananas the Clown to arrive at their birthday party.

This is what homeschooling through high school has brought me? A nice, steady job...as a clown?! The thought is enough to drive anyone to a military academy. Perhaps upon seeing me at work, the first instinct of today's up-wardly mobile student is to run as fast as he can toward the nearest school. But with a second look he can see that my plans for the future are actually bigger than my 17-inch shoes, and I can honestly tell him that if I had to relive my high school years I would homeschool again.

I imagine there's one thing running through your head right now: "How in the world did this guy become a clown?" Or more likely: "What possessed this guy to be-come a clown?" I'm anxious to tell you, but I haven't even introduced myself yet. My name is Jonathan Rockett. My friends at college call me "Johnny Rocket." I have two older brothers and one younger brother. The Lord did not feel it was necessary to smite me with sisters. I was born in San Jose, California, but my family moved to Texas while I was still a baby without even asking me how I felt about it. We currently live in Maryland. I was taught at home from kindergarten through twelfth grade. I loved Bob Jones literature, I hated Saxon math, and I never want to see the A Beka Biology videos again. I like theater, poetry, antiques, puppies, gummie bears, figure skating, and Beethoven's "Pathetique Opus 13," which I'm listening to as I write this. (Sorry, for a minute there I thought I was writing an ad for the personals section of the paper.)

I wish I could promise you that homeschooling through high school will always be fun and easy, but it won't. I wish I could guarantee that you'll never want to go to school, but I can't. I wish I could tell you that you will live in perpetual bliss with your siblings and parents, but you probably won't. (Heaven knows I didn't.) All I can do is tell you my story and hope that it excites and encourages you. I believe homeschool can provide a superior education and a whole lot more, but you have to work for it. They say that the proof is in the pudding. I'm just glad I like pudding.

Finally, we're back to the clown story:

...It all started when I was twelve years old. I wanted to find a way to make money outside of the lawn-mowing circuit. I had taught myself to juggle when I was about seven years old, and I've always been fascinated by magic tricks. I loved children and loved to be silly even more. It seemed clowning was my density, I mean my destiny. One bright, spring morning, engrossed in the planning of my new business, I saw an ad in the paper. It was for the American Red Cross Safety Clown Troupe which performed skits about safety for preschoolers. I joined.

A couple of years slid by and my career as a "Safety Clown" gave way to my growing birthday party business. I bought a unicycle and attended a ten-week clown school—Bananas the Clown had become my second identity. Since then I've performed as a guest clown with the Ringling Bros. and Barnum & Bailey Circus and also with the Clyde Beatty Cole Bros. Circus. I'm a Member of Clowns of America International and among those in my Rolodex whom I call my friends are names such as Kenzo, Chuckles, Jingles, and Ronald McDonald. (I know. It frightens me too.)

Clowning has been a large and enjoyable part of my life, and I am continuing to clown my way through college. Uh, outside the classroom of course. I certainly don't intend to make a career of it, and I'm not sure how long I'll continue, but it has been an exciting source of invaluable experience. I owe the opportunity to homeschool, which gave me the time, flexibility, and self-motivation to shape my interests and talents into a business. I learned that all the world does love a clown, but, more importantly, I was being prepared for what was to come.

I don't know if I realized it while it was happening, but during my early school years I was developing a deep love for the performing arts. My favorite field trips were always the ones involving performances. The gears whirred in my little pea brain and I thought, "I could do that." Well, maybe I didn't decide so deliberately, but my every waking thought, and even sleeping thoughts, began to revolve around writing scripts, collecting dress-up clothes, and entertaining audiences. After I began clown-

ing, I joined the church drama team and got a taste of acting. I loved every second of it. They appointed me Prop
Manager and eventually I began directing. It felt strange as
a fifteen-year-old directing adults. (Okay. I admit it. I got a
huge kick out of it.)

During my sophomore year our homeschool oversight
program decided to produce a musical. It was a big step,
but we worked hard, and before we knew it, we had performed the *Sound of Music*. It was an enormous success
for a homeschool group. The next year, we produced *The
Music Man* with even greater success. My senior year saw
Fiddler on the Roof bring sellout crowds and standing ova-
tions to Smith Theatre at Howard Community College.
That small group of homeschoolers had gained quite a
reputation.

It was during my junior year that I developed a vision
of what I believe God has called me to do. Now I can't see
myself doing anything else. I'm excited and eager to fulfill
the plan He has for me. I am also determined because it
means jumping head first into an industry that has turned
its back on God and thrown away all moral ties. But I'm
getting ahead of myself.

I believe that God is a very creative and artistic God.
Every time I step outside I am overwhelmed by that fact.
The Bible says that we are made in God's image. Maybe
part of what that means is that we have creative, artistic
abilities like Him. I also believe that it gives God great joy
when we use those abilities. I like to think of our art as
God's refrigerator art. Beaming with pride, He says, "See
what my kid did?" He has given us this thing we call art to
amuse us, to cheer us, to shake us, to appall us, and to
nourish our souls.

I asked myself, "What is, arguably, the most powerful artistic medium today?" The answer was film. Movies have the potential to affect millions of people. But there is a serious problem. Filmmakers have misused that power and all but destroyed our country's moral foundation. I'm excited by the way movies shape our society, but I am saddened by the shape that has emerged. Christians should be ashamed to admit that we have all but given the industry away and allowed others to give Hollywood its identity: The Armpit of Western Civilization. You see, Christians often forget that they have more reason to make movies than anyone else, that they can glorify the Creator of the universe. In fact, many Christians avoid movies altogether and for good reason. This should not be. I believe that someone must roll up his sleeves and create a new standard of excellence for film.

My mission is twofold: First, I want to remind the world where art comes from by using film's influence in a positive way, demonstrating that superior artistic quality goes hand in hand with superior moral quality. Second, I want to be a witness to those I come in contact with in the industry. How many times have we shaken our heads and remarked that Hollywood desperately needs Jesus? Well, somebody has got to introduce them. This is my vision.

The clock on my desk reads 7:51 p.m. My eyes have found an interest in everything in my room but the paper in front of me. My calendar says I have 54 days until I begin my junior year in college. It seems like I just got out of high school last week. They tell me the next two years will go even faster. I'm attending Hillsdale College in southern Michigan and pursuing a degree in theater and speech. I'm so thankful for this opportunity and excited about where I am. The faculty and the students are top notch, and the opportunities are endless. I'm very involved with the theater department, write for Hillsdale's literary magazine, do volunteer work, and I recently started a band.

Before my return home, I prepared myself for the laughter that I thought might come from my family and friends when they heard that one of my extracurricular activities was ballet. I'm glad I prepared myself. Boy, did they laugh. I started taking ballet because I thought the exposure would improve my awareness of body movement while acting. But now I'm continuing because I discovered that I love dancing, despite the tutu jokes.

This summer I will perform a short play at the Edinburgh Fringe Festival in Edinburgh, Scotland. It is a once-in-a-lifetime opportunity to perform at one of the world's largest theater festivals, and the eight of us who are performing the play are extremely excited. As each new door is opened and more opportunities are discovered, my confidence in God's plan and His active participation in it grows stronger.

The main thing I noticed about my transition from homeschool to college life was that I hardly noticed it. The transition happens whether you want it to or not. Even if I could pick apart and analyze my transition, whatever that is, it would be useless because every college freshmen has things to learn and adjustments to make whether he homeschooled or not, and everyone makes them differently. Remember that none of the other freshmen have been to college before either.

Perhaps I should qualify. I do feel that, academically, I was better prepared for college than the average student. There are two reasons: The first is that my brilliant teacher (mom) put a huge emphasis on writing and communication skills during my years at home. The ability to write well and communicate your thoughts is the foundation of success in school and life after, even if you are not an English major. I have seen professors of history or psychology grade papers tougher than many English professors do. Learning to write well will actually help you in every subject.

Secondly, I felt prepared for college because homeschool gave me a strong work ethic, independence, and a love for learning. I'll insert some honesty here—sure, I didn't wake up every morning with a burning desire to do math problems, but I did enjoy learning about things that interested me. It is true that there are plenty of homeschool students who have poor work ethics and aren't ready for college. Conversely, there are plenty of non-homeschooled students who have great attitudes and are ready for the most challenging of schools. Homeschooling does not automatically make you a strong student; just like anything else, it takes work. But I believe that homeschool provides an atmosphere more conducive to the development of a love for learning.

Before I left for college I wondered how my worldview would be challenged. I was excited to take Summit Ministries' *Understanding the Times* course with other high schoolers in my homeschool group and to attend the two-week Summit leadership conference. I knew my biblical Christian worldview was true, but Summit helped me learn why it is true and how to defend it.

Hillsdale is not a Christian school, but it is very conservative and founded on Judeo-Christian principles. It has a strong core of Christian students and faculty, and Intervarsity Christian Fellowship is the largest student organization. Even so, I have still found opposition to my faith. I am friends with Catholics, Jews, atheists, and agnostics, and we spend long hours debating. In doing so, my faith is strengthened.

Let me share an example with you. I was in my physics class one afternoon and we were discussing Quantum Theory. The professor told the class that, "...anything can happen in the universe. For example, a man could walk through a wall. The reason things like this don't happen though is because the probability is so small. But there is always a probability of it, small as it may be." The conversation continued and he eventually made the point that miracles don't exist, or at least they are not caused by supernatural means. Instead, they are just occurrences that we rarely see due to their small probability. I thought to myself, "A statement like that only comes from a man whose livelihood depends on his ability to explain the physical world. He simply can't accept information outside of his scientific powers." I raised my hand, and the remaining half of class was spent in a debate between the professor, a nuclear physicist whose former partner recently won a Nobel Prize, and me, a freshman who was taught at home. The debate gathered momentum... He said something like, "All I'm saying is that there is no way to know if miracles or anything supernatural exists, because we can't prove it or test it. We just can't know." So I said something like, "In the textbook you've chosen for this class, there is a whole chapter devoted to extra-terrestrials and how we might communicate with them. Show me the evidence for that. By definition, the <u>super</u>natural is outside the realm of science. Of course it can't be proven or tested by science. But it also can't be disproven by science. How do you know that there isn't some nonphysical part of your identity, for example a soul, that allows you to experience and comprehend the supernatural? Faith is the evidence of things unseen." That was the end of class. Out in the hall I learned who all of the Christians in the class were because they came up to me and expressed their support for the way I spoke up. By the way, although we disagree on metaphysical issues, he was the best teacher I've ever had and one of my favorites.

"I have a dream." This statement is far from original, but that fact does not rob me of any of the simple strength it carries. I have a dream to show a world full of people that there is a God who created them, who loves them, and who wants to share eternity with them. I have a dream to create movies that raise important questions in the minds of those who will watch. Many people have dreams, but my dream comes from the knowledge that I have a purpose in the universe God has created. I said that if I had to relive my high school years I would homeschool again. Our decision to homeschool not only helped me find my dream, but it also gave me a strong foundation for attaining it. And all dreams that are in the Father's plan will come true.

Don't look now, but there is an entire generation of homeschoolers who are growing up and leaving home in pursuit of their unique purpose in the Father's plan. They are spilling into society, and the world will find it hard to ignore them. It is often asked whether homeschoolers are ready for the "real" world, but homeschoolers are asking whether the "real" world is ready for them. As a homeschooler, you have been blessed with a wonderful opportunity. Never take for granted the things that you can accomplish. Even a clown has a place in God's Kingdom.

*Donielle Mohs lives with her
family in Moline, Illinois.*

CHAPTER 3

In the middle of the fourth grade, my family moved, and I had to change schools. My experiences at that new Christian school were devastating.

Before the move, I had been one of the top students in my class. But for some reason, the new teacher got a bad first impression of me, probably because I was painfully shy, and decided to tag me as a liar.

As the year progressed, the teacher's attitude toward me rubbed off on my classmates, which, in turn, rubbed off on me. I was no longer the top student in my class, and by the fifth grade I had become what my fourth grade teacher had tagged me, a liar. I would lie about my homework, telling my teacher I had done it but that I had forgotten to bring it, and telling my parents that I hadn't been assigned homework. My grades fell rapidly. I would come home from school crying.

But God was watching over me. Nearly a year later, my dad read something about homeschooling. He asked my mother whether she would consider trying it. She was wary at first but willing to give it some prayer to see if God wanted her to do it. God answered. We tried it that summer, and we loved it.

My grades improved miraculously, as well as my relationships with my parents and brothers and sisters. Most importantly, through daily contact with the godly examples and instructions of my parents, my relationship with the heavenly Father grew stronger. I no longer spent most of my day under the influence of lukewarm Christian or even non-Christian peers.

Through godly homeschooling, I have a solid foundation in Christ rather than a weak and crumbling foundation in my friends or in myself. Because of this strong foundation of faith, I am able to stand against "the wiles of the devil." I am not easily influenced by what others think of me because I know that God's opinion of me is the only important one.

Without homeschooling, I know I would have continued down the path of a liar. My grades would have kept dropping, and I would have become more and more miserable. I thank God daily that He pulled me out of such a degrading situation. He truly is my provider and my salvation.

THE NEW PROJECT

I walked into the classroom, huffing and puffing after the long walk from the car. "Whew," I thought to myself. "I'm sure glad this is just an art class. My hair is a mess from all that wind." I watched as my classmates settled themselves into their seats and took out their pencils and sketchbooks.

"We start a new project today," I said to one of the students sitting next to me. "I wonder what it will be. Have any idea?"

"I don't know," she replied while doodling on her paper, "but I sure hope it's fun."

A hush fell over the classroom. Everyone focused their attention on the professor entering the room. "Today we're going to work on figure drawing," she said. "A model is coming in to pose for us—a nude. We must learn to draw uncovered figures so that our covered figures will be more accurate. Try to make the drawings as detailed as you can."

I felt sick to my stomach. All my excitement about this new project flew out of the window. I couldn't believe that the professor was going to make me do that. A nude model? This was the very reason I'd almost decided not to major in art in the first place. I felt angry. I wanted to get up and walk out. Yet I felt confused. After all, God made the human body. He made us in his own image, the crown of his creation.

I will praise Thee; for I am fearfully and wonderfully made: marvelous are Thy works; and that my soul knoweth right well (Psalm 139:14).

The human figure is beautiful. It's one of the more interesting things to draw. In fact, some of my favorite paintings are of human figures. Another verse came to mind.

Finally, brethren, whatsoever things are true, honest, just, pure, lovely, of good report; if there be any virtue, and if there be any praise, think on these things (Philippians 4:8).

I had no problem with drawing a human figure, clothed. But the idea of drawing nude figures disturbed me. I didn't want to participate in something impure, unjust, or without virtue. I thought about the scene in the Garden of Eden, when Adam and Eve had disobeyed God by eating the forbidden fruit. They felt ashamed because of their nakedness. They tried to cover themselves with fig leaves in order to hide themselves and their shame from each other and from God. But wimpy leaves don't cover much. So God, in His kindness, clothed them with animal skins. God did not give Adam and Eve clothes because they got cold or because He thought it might look good; He clothed them because of their sin and shame.

This assignment was shocking. Did anybody else feel the same way? I hoped one other person might object as much as I did. I looked around at my classmates to see their reactions. It was worse than I expected. They all seemed perfectly fine with it. In fact, one guy behind me remarked, "All right! Finally this class is worthwhile. A nude model! Cool!" I found this utterly revolting. I prayed that God would give me the strength to take a stand. In the world of art, nudity is expected, even praised. For me, nudity is nakedness. Period.

As my fellow students prepared for the drawing session, I approached the professor. The moment I left my seat, I felt the devil trying to sidetrack me, making excuses for why I should just be quiet, that it really wouldn't be all that bad to draw a naked person. With each shaky footstep, I struggled. How could I say what I needed to say without offending the professor? How could I be humble while at the same time standing up for my convictions?

In our homeschool, I had been taught that respect for anyone in authority is very important. I was also taught that, if the authority acted contrary to God's Word, then I should oppose the wrong, gently and firmly. I took a deep breath as I stood before my professor. "Could I talk to you about this assignment?"

She turned and looked at me. "Yes?"

My heart pounded. "Oh boy, what was I going to say now?"

"Well, I'm not comfortable with drawing nude models. Is there any way I can have my sisters pose as clothed models instead?"

She didn't seem very happy about that. My mind was racing, trying to think of a way to show her that I was willing and eager to draw a human figure, just not nude.

After a few moments of excruciating silence, she sighed and adamantly said that I still needed to attend class. My heart sank. I knew I had lost the battle. I wanted to cry out to God and say, "Why aren't you helping me? I don't want to do something improper, but I also don't want to get an F in this class. What do you expect me to do?" Then, to my surprise, the professor went on. "Well, I suppose that I could have a nude wear a bodysuit or a swimsuit for the first part of each session. At the end of the first half, I will excuse you from the rest of the class."

"Yes!" I thought to myself, "That's a perfect solution! God certainly knows how to answer prayer!" As I went back to my seat, I rejoiced in the knowledge that God was watching over me. He knows how much I enjoy drawing His creation and did not want to miss out on the opportunity to draw His highest creation, the human.

During the first part of the class each day, I enjoyed myself thoroughly. The model was intriguing. She held what sometimes must have been very fatiguing poses. I loved drawing the smooth curves, the hard lines, and the subtle shading. When the time came for me to leave, I quietly gathered my things and walked out.

At first, I felt really conspicuous. "Why am I doing this? People will think I'm some kind of weirdo or fanatic." Sometimes my peers would give me strange looks. Sometimes I felt they thought I was a "goody two-shoes." Even the model glared at me once or twice. I think she figured out that I thought what she was doing was wrong. I prayed that God would convict her and bring someone into her life someday to witness to her and show her how immodest it was to expose herself.

As the days wore on, each time I picked up my things, I could feel God smiling down on me. I began to have more of a spring in my step as I walked out of the classroom and down the hall. I was not afraid of what my peers thought of me. I knew in my heart that my actions were pleasing in God's eyes, and that His eyes are the only ones that count.

God blessed my efforts to remain pure by allowing me to get an A in the class. He enabled me to plant a seed of doubt in the minds of some of my classmates about their lifestyles. My professor could also see in the way I chose to live, a difference from most people.

As I look back on the situation, I know my courage in the face of the world's accusing stare came from my God-centered home education. Many college classmates of mine have expressed uncertainty and fear about standing against something they feel or even know is wrong. I believe that this anxiety is caused by a weak foundation in the truths of God's Word. Only God's strength is sufficient to overcome all things. As Philippians says, "I can do all things through Christ who strengthens me."

HOTHOUSE TRANSPLANTS

Jennie (Ethell) Chancey
graduated summa cum lauda from
King College in 1994.

CHAPTER 4

BEYOND HOME HIGH SCHOOL

Gravely shaking her head at my mother, the reporter glanced at me as I sat working on an assignment. "But Mrs. Ethell," she murmured, "how will your daughter ever learn to cope in the real world? What if she decides to go to college without ever learning to walk down the halls at a real school or open a locker?" Mom stifled a chuckle. The interviewer had never heard of homeschooling, and this was her first exposure to kids who spent their days hitting the books at a dining room table or on the couch. By the time the reporter left that day, she was fairly sure we'd have a fabulous homeschool experience up through junior high. However, after eighth grade, we'd just have to go back to high school or face serious problems down the road.

Nearly seven years and two graduation ceremonies later, I wonder what ever became of that reporter and what she thinks of homeschooling now. My appreciation for it has only grown since my entry into the "real world," and I would never trade those years at home for anything. Without homeschooling, I would have missed the most educational, challenging, and fun experiences of my life.

I'm happy to report I didn't have any problems walking down the halls at college or opening and closing lockers. Nor did I get lost looking for my dorm room or have to ask someone what a lab science was. And I didn't hide in the closet when it came time to "socialize." The greatest struggle for me was not in signing up for classes or learning the campus layout, but in conquering homesickness. I missed my family terribly. There wasn't anything wrong with my social life—I found plenty of friends both in and out of class—but it was hard living five hours away from the ones I loved the most. For the first several weeks, my parents kindly put up with daily collect calls. Eventually, I cut down on the phone addiction and took to letter writing. And, although I never really stopped missing my family, I grew accustomed to living away from them and became more and more excited about all the things I could study at school.

College presented an atmosphere of learning for which I felt fully prepared. The Christian liberal arts school I attended had small classes and very accessible professors. It was like an extension of homeschooling. I loved getting into the materials, participating in class discussions, and writing papers. I even learned to love biology, though I admit cutting into frogs and pigs has never been a special love of mine! My professors expressed amazement when they discovered I had been homeschooled. I was the first of "my kind" at the school, so I was put under the microscope in many ways. Sometimes I got talked about in the third person: "Her parents seem to have taught her good study habits. How do you suppose she cultivated those without any homework?" At first I was puzzled. Study habits? What are those? Why wouldn't every student study? It just made sense to get the work done to make room for the extras. If I was going to get a diploma after four years, I wanted it to mean something. (And colleges are looking for students who want more than a piece of paper for all their work. When I worked in media relations at Home School Legal Defense, I got calls from colleges and universities looking for homeschoolers. Professors all across the U.S. bemoan the lack of motivated students, and they've seen a rekindling of the love of learning in homeschoolers.)

But my college experience provided more than interesting topics to study and people to meet. It also provided a window on other worldviews. According to many of my peers, my beliefs were nonsensical. To them, homeschooling meant sticking my head in the sand and refusing to acknowledge "reality." Sometimes that bothered me. Homeschooling had taught me to think on my own, evaluate other worldviews, and assess them using absolute standards. It puzzled me that my peers were often tied into the status quo, following along with whatever sounded right at the time—all in the name of "free thinking!" This was their reality. But I couldn't buy that. Sometimes I found it more comfortable to talk about "reality" with professors than fellow students.

There was a bit of culture shock involved in going to college. Even in a Christian school, there were students who hated church and didn't believe in a biblical worldview. My parents had prepared me for this fact, but I had no idea what a challenge it would be to defend my beliefs. That was an education in itself. It drove me back to the Scriptures time and time again.

I think it was the "old-fashioned" ideals my parents gave me through homeschooling that helped me through college more than anything else. One thing mom and dad taught each of their kids is that our lives are not our own—they are God's. We do not go to college or get a job to satisfy our own desires, but to follow the will of Christ. We were also encouraged to think "generationally" rather than to focus on our own lives. That meant keeping the future of our children (Lord willing, there will be many!) in mind when making plans. What would they need to learn? How could we prepare to raise the next generation? Mom said, "It isn't about what you want. It's about God's calling on your life." There were times this thought overwhelmed me. Couldn't I be just a little selfish? After all, I was the one doing all the learning at college! I should take classes I wanted! But obeying mom and dad actually led me to the writing and literature classes I was dying to take, along with healthy doses of history, science, and math. It wasn't always easy to think beyond myself, but my parents helped me expand my focus. Even 300 miles away, they were still my greatest support system and help. That's a "family value" that works!

According to many people, my beliefs in the importance of family life, culture, and biblical law are "unrealistic in today's modern world." I can't count how many times I've been told I was born in the wrong century! Amazingly enough, my parents didn't weep, wail, gnash their teeth, or pull their hair out when they heard they'd "failed" to teach me "modern" thought. Instead, they increased their efforts to teach my younger siblings in the same fashion. That has yielded fantastic results. Instead of becoming "social misfits," my brother and sister have gone on to great success. David was offered a job with a computer consulting firm before he graduated from college. His bosses say his hands-on experiences with programming and computer hardware during his homeschool years were more valuable than his degree in physics and computer science! David didn't just read about the work he wanted to do; he did it.

My sister, Julie, also entered the working world ahead of her time. Her interest in art led her to study graphic design. Going beyond the basics, she designed and laid out a book which was published the summer she turned 16. A second book was in the works by the time she graduated from home high school. Her work has been highly praised, but she blushes at any applause. "I'm just doing what I like to do!" she says. This is how it should be. Instead of avoiding the real world, homeschooled kids really live and function in it.

It is important to get to the bottom of the whole "real world" talk. What is the "real world?" How do we prepare for its challenges? We live in a crumbling culture and an ailing society. How can homeschooling prepare us to hold on to our worldview when the "real world" thinks that view is backwards and archaic? Personally, I believe it is the only way our society will survive. My job at Home School Legal Defense gave me daily opportunities to talk with the media about homeschooling. Dozens of reporters told me that homeschooling made sense on more than an academic level. They talked about school children shooting each other, doing drugs, getting pregnant, and dropping out of school, and they wondered where it will end. Some people are finally waking up to the tragic conditions created by the public school system. It is a system that has failed, and more people than ever before want out. This is where "old-fashioned" values come in.

More than anything else, these values helped me to logically think through opposing viewpoints both in and out of the classroom. If you have a strong foundation on which to build, it is unlikely that you'll topple easily. My parents made sure each of us not only knew what we believed, but why we believed it. They helped us personalize their beliefs and make them our own. These beliefs obviously included our faith in Christ and His salvation, but they also encompassed thoughts on the church, civil government, education, gender roles, marriage, family, child-rearing, and more. My beliefs earned me some odd looks at college and some hurtful comments. At times I wondered if I really was a nut whose parents were out of touch with the world! Again, it was constant contact with my family that kept me thinking, evaluating, and debating. I also found some good mentoring relationships with a couple of professors. They were intrigued by the way my family thought and wanted to know more about it.

For two years after college I worked in media and public relations at the Home School Legal Defense Association. I had opportunities to talk with every major television network and the largest newspapers in the world. Reporters called daily to ask why homeschooling was growing and how well homeschoolers did in the "real world." One of the most rewarding parts of my job was to see the shift away from stereotyping homeschoolers as backwoods religious nuts and the drive toward portraying them more accurately as concerned, loving parents who want the best for their children. Slowly but surely, the media have woken up to the excellent trend that homeschooling represents. I'm honored to have played even a small part in this shift.

But I know I couldn't have handled the job if I hadn't been taught at home. I was able to share "hands-on" knowledge with the reporters who called, often giving them the added angle they need. Because my parents nur-

tured my interests in communications and writing, I didn't
find it difficult to jump right into my job and use those
skills. I had also worked for my father during high school
and between semesters at college, assisting him with research, interviews, and public relations.

Although my college study helped hone my writing
skills, it was really homeschool "socializing" that helped
me get where I am today! Mom and dad encouraged us to
talk with both older and younger people, and, although I
was initially shy, I quickly felt at home getting to know
people. It didn't matter that my favorite penpal was 65 or
that I loved playing with little children—indeed, it helped.
I had no problems relating to the people who called me on
the job to get statistics, set up interviews, or just chat about
education. Homeschooling prepared me for this. And
every day I am more convinced of the effectiveness of
teaching kids at home. The testimonies I read in the daily
papers, the shows I hear on radio, and the personal calls I
continue to receive back this up. It's not just about good
grades, it's about life.

Homeschoolers represent a growing part of the popu-
lation that is concerned with America's slide away from its
moral base. They also make up the most active part. There
is hope in this, but only if parents continue to teach a
"radical" (by the world's standards) worldview and
homeschooled students continue to hold to that view in
spite of cultural opposition. My parents always believed
the focus of homeschooling should not be to create students who can spout off lists of facts and figures, but to
bring up responsible Christian men and women who can
think, reason, and apply Christian principles to daily life.
More than anything, this focus has given me the greatest
sense of responsibility in my life. It places a weight on me

to think carefully before I make a decision or judgment. Instead of being a burden, though, this is a tremendous help in the "real world."

If I had it to do all over again, I would definitely want to be taught at home. I retired from my job in 1996 to become a full-time wife. And I know that when I have children, Lord willing, I will teach all of them at home. Perhaps by the time my son reaches high school age, he'll be able to tell reporters how homeschooling went from *a* choice in education to *the* choice in education. I hope so. Homeschooling made my transition into the "real world" smooth and straightforward. Thanks, Mom and Dad!

HOTHOUSE TRANSPLANTS

Matt Chancey lives with his wife Jennie and brand new son in Virginia, and plans to attend law school through the University of London External Degree Program.

CHAPTER 5

HOMESCHOOLING: GOD'S VEHICLE FOR MY LIFE

On a cold winter day in 1987, a small hearing room in Montgomery swelled with over 500 homeschoolers. Alabama was one of the few states in the deep South which had not legalized homeschooling. That day, these peculiar southerners were attempting to change all of that. To help the families present, a young attorney from a small legal office in Washington D.C. flew in to give testimony on the merits of homeschooling.

Chris Klicka was in his mid-twenties but had already been involved in some of the most important homeschool cases in the country. He had been called to defend American families, and now he stood before the leaders of Alabama, surrounded by their education union minions, to do just that.

Representing the opposition was the most powerful Alabamian in education: Paul Hubbert. He pulled all the strings in the legislature, and now he faced a young hotshot attorney from Washington D.C.—one with a Milwaukee accent of all things! Mr. Hubbert appealed to the "carpet bagger" defense: "What right do these outsiders have to come in here and tell us how to run our government?" The young attorney coolly made his case, present-

41

ing all the arguments, statistics, and facts. An 11-year-old boy watched in admiration as the young attorney testified.

The situation didn't look good for the homeschoolers. Every time a vote was called, Paul Hubbert stared at the intimidated legislators, who rose from their seats and left the room, disallowing the necessary quorum to initiate a vote. Frustration filled the young attorney's eyes as the day came to a close.

Alabama homeschoolers lost that day, but the day's work had not been in vain. Eight years later, Paul Hubbert ran for governor on a pro-homeschool platform. A powerful precedent had been set in 1986, and no one could take away the respect that Chris Klicka and the Home School Legal Defense Association had earned from the testimony given on that cold, wintry day. The eleven-year-old boy in the hearing room didn't forget, either.

When my parents began homeschooling me, I had the initial reaction of any arrogant, selfish ten year old: "What? You can't be serious! What will I do about my social life? My parents don't know anything. 'Wisdom was born with me!'" Little did I know just how foolish I was and how much my thinking process had been altered by the Christian school (yes, you read rightly) which I had attended for five years. To me, homeschooling was like piano lessons; I hated it and I knew that I would always hate it.

The Christian school I had attended was not a "values-neutral" establishment. Its problem was that it was like most other schools—it was institutional, not interpersonal. A Christian school, like many other parochial institutions, in seeking to educate its students in all godliness, actually can take away the best years of children's lives by separating them from the people who love them the most.

42

To make up for the emptiness left by the lack of a patriarchal family unity, a rugged individualism soon became the substitute, sowing the seeds of rebellion which bloomed in me, as in most other teenagers.

However, by the grace of God, I later began to see my error, realizing I would one day leave home forever and never experience the same covenantal relationship I had when I was with my family. The times were changing. I was growing up and trying to come to grips with who I was and what I was here for. In short, I had my first mid-life crisis when I was 14!

As far back as I can remember, I have wanted to fly. Flying had been in my family's blood from the time my grandfathers flew P-51s and 47s during W.W.II. I had the opportunity to take flying lessons for a short time in a small Cessna, which I loved. I decided I wanted to join the Air Force, and I focused on applying to the Air Force Academy. Everything was going great; then the bomb fell.

When Bill Clinton was elected, I was crushed. All my life, I had wanted to serve my country in the armed forces, and suddenly I was forced to consider taking orders from a man for whom I had virtually no respect. Since homeschooling had taught me the merits of planning ahead, I had already considered alternatives if "Slick Willy" was to be elected.

Politics had always interested me, but I had thought I'd only get involved after retirement. I had given little thought to devoting my life to such an unpopular calling. But I had to admit I liked the action. So, I began a hobby which would turn out to be a full time calling.

THE POLITICS OF POLITICS

Politicians love youth. To have young people will-
ingly pull for your campaign is about as frequent as a snow
storm in Tahiti. I knew the opportunities to really make a
difference in local politics were out there; I just didn't
know whom to ask. I was given the name of the chairman
for the Republican party in my county, and I called him.
He was very excited that I was interested in helping "The
Cause" and was even more excited that I had several
friends who were also willing to help. The time was January, 1994, and my life was about to speed up considerably.

Being homeschooled was chiefly responsible for what
happened to me during the next year. I was able to meet
the most interesting people, simply because I had the time
flexibility to do it. I was able to meet and talk with my
governor, attorney general, supreme court justice, and numerous representatives and senators. They were equally
pleased that a young person was interested in their plat-
forms. I would simply walk up to a candidate whose position I agreed with and say something to the effect of,
"Here I am! Do with me as you will." Believe me, they
did. Several late nights, door-to-door visits, endless sign
distributions, political rallies, and ten months later, Alabama Republicans had one of the most successful races in
its history. I was able to take part in some of the most
thrilling races—races in which the winner was decided by
less than 70 votes, and all because I was not tied down by
an 8 a.m. to 3 p.m., five-day-a-week school schedule.

In addition to political work, I started my first year of
college. While being homeschooled, I took courses at a local junior college and was able to accumulate 30 hours of
credits before I finished "high school." While in college, I

went to school in the morning, and worked in politics in the afternoon.

INTERN PROGRAM IS STARTED AT HSLDA

After the election "high" had subsided, I heard about an internship program starting at the Home School Legal Defense Association. HSLDA advertised a program which would center around political activity, stirring my interest immediately.

I called HSLDA's office in early January and talked to a young woman, Jennie Ethell, who shortly after faxed me an intern packet. I hastily filled out my application and mailed it off. A couple of weeks later, I was informed that I had missed the first deadline and that my application would be held over until June when the next round of interns would be selected. I was starting my third quarter at a local junior college, so I hit the books as the months flew by. Homeschooling had done a good job preparing me for the college life, and by the time I finished my first full year, I had a 3.98 GPA.

In May, Jennie called to tell me I had been selected for a phone interview.

A week later, I was interviewed by several members of HSLDA's staff, including the attorney who had come down to Alabama almost ten years prior to testify for homeschoolers—Chris Klicka.

The interview went very well (despite me being a nervous wreck!), and I was informed the same day that I had been accepted. I was "tickled pink," as an Alabamian would say.

I had a month to prepare, and I busily worked to get ready for what was to be the best six months of my life.

TO VIRGINIA!

The traffic on the Fourth of July was rough as I worked my way up the east coast toward a little town of which I had never heard. Purcellville, a town with only one stop light, contained one of the most respected organizations in the nation, the Home School Legal Defense Association. I did not know what I was in for, but I knew I would love it!

When I arrived, I met Jennie, and she took me to the intern house I was to live in for the next six months of my life. When I had first talked to Jennie on the phone in January, I pictured her as a married woman in her mid-thirties. Instead, I met a single, attractive young woman in her early twenties. Little did I know what the Lord had in store for me.

HSLDA was fantastic. I had no idea of all that this small organization did for homeschoolers. Mass mailings, state and federal legislation alerts, leaders' mailings, radio programs, lobbying days, and endless legal battles, etc., all accomplished with a staff of under fifty people. I was excited to be a part of this team and would have been content to make photocopies all day long if that was asked of me.

The six months I spent at HSLDA were terrific. My experience in that little town of Purcellville was better than the best political science course at the largest university. I worked in the National Center for Home Education division of HSLDA. I was placed in the office next to Chris Klicka and Andrea, his assistant. For the entire six months, I stayed at the same desk taking in all of the politics like a dry sponge soaking up water.

AFTER HOURS..

In the meantime, I began seriously considering initiating a courtship with Jennie. I had never read a single book on courtship, but I pretty much knew what I needed to do. Homeschooling had shown me the importance of the family and the covenant it represented. I knew that if I courted Jennie, I would have to secure the blessings of my family as well as hers. Jennie said it best when she described my courtship to her as "Matt's courting of my whole family." When I was sure Jennie was the woman God made for me, I wrote my dad a lengthy letter asking for his blessing and advice. He responded in the affirmative, and sent a letter of recommendation which I was to give to Jennie's folks when I asked them for a courtship—that's right, I said, "Asked her parents."

When I drove over to Jennie's parents' home on a beautiful fall day in October, Jennie had no idea what I was doing. I ate dinner with her parents and "popped the question" without her knowing anything about it. This, of course, was contrary to the individualistic concept of dating which treats parents as unwelcome observers. Mr. and Mrs. Ethell both looked at each other and then at me. "Well, we kind of figured that was the reason you were coming over," said Mr. Ethell, "and you have our permission. Have you talked to Jennie about this yet?" "No, sir," I said. "Well, you are going to have to tell her now," he replied. "When do you plan to do this?" I looked at my watch, calculating how long it would take me to get to Jennie's place. "In about forty-five minutes," I responded.

Jennie had gone out to eat with her roommate, so I waited about 45 minutes for her to return. When she arrived back at her apartment, I asked her if she wanted to go out again. She answered in the affirmative, and we went

47

out to Shoney's for coffee. We were there from 9:00 p.m. until 1:00 a.m. As we sat together in the quiet restaurant sipping half-burnt coffee, I asked Jennie for her permission to court. Speechless, she nodded a "Yes."

NOW WHAT?

Well, I had done it. By modern standards, I had broken every rule in the book. I was not even twenty years old; Jennie was 23. I had only 1 1/2 years of college under my belt; Jennie had graduated with honors a year before. I had no real job, and was not sure what I was going to do in December when my internship ended; Jennie had a job and security. I had initiated the relationship in a totally old-fashioned way. I, personally, was an incredibly old-fashioned guy who frowned on everything contemporary from big screen TV to day care. In short, I was unacceptable by all modern standards. What right did I have to initiate any sort of relationship with anybody? God knew better.

Shortly after our courtship began, Andrea announced she would be leaving in December to go home to Wyoming and get married. This meant Chris Klicka would need a new assistant to take over. You can imagine my shock when Chris offered me the job. The Lord's plan began to unfold. I could see everything so clearly. Everything that had happened in my life, from the time I met Chris in 1986, to the political involvement in my teens, to the internship at HSLDA, and even to the place I sat right outside of Chris's office to observe how everything worked at the National Center, was preparing me for the day and the time I would meet the woman God had prepared for me. Homeschooling made it all possible. If I had not been homeschooled, I would never have received the political experience I did. It is likely I never would have developed an interest in politics. I would have been unable to get an internship with HSLDA even if I had an interest in politics, and most importantly, I never would have met Jennie.

I am now working full-time at the National Center for Home Education as Chris Klicka's assistant, and Jennie and I were married on June 22, 1996. Our first child, John Nathan, was born on May 30, 1997.

Some of you reading this story might just be starting your homeschooling experience. You are probably like I was when I started. When you get frustrated, remember my story. I'm no one special. There is no particular reason why God orchestrated the beautiful plan in my life the way He did except for His love and His purposes. God brought all of the wonderful things about in my life by using homeschooling as the primary vehicle. Enjoy your family. They are your best friends. You never know if you will be called away on an internship or other work experience and never return home. Homeschooling allowed me to live the "family values" so many people just like to talk about. You can look ahead in excitement to what God is going to reveal to you as you continue to grow up at home.

*Matt and Jennie Chancey were
married on June 22, 1996.*

HOTHOUSE TRANSPLANTS

RaeAnn Hering lives with her family in Fountain Valley, California.

CHAPTER 6

I'm in ninth grade, standing outside the locker room waiting for the P.E. teacher to come out and start the class. I've just come from four years of homeschool. I convinced my parents to allow me to go to public high school. I thought I'd like to give it a try and see for myself. I've seen. Now it's late November and the novelty has worn off. I have already attained straight A's and a few friends. I'm in choir and have just been accepted into the school band. So why do I feel like I'm wasting my time? There's no challenge. As long as I show up to classes with my finished assigned homework, I get an A. And it seems to take the teacher half of each period to get the class settled. I'm here for eight hours, go home and do three more hours of homework, then eat and go to bed. School is my life, and yet I feel unproductive. My mind keeps wandering back to homeschooling. I would feel so fulfilled if I could study something I enjoy more. Homeschooling had allowed me to do this. Sure, public school has its advantages, but I can't find enough to outweigh the advantages I can think of in homeschool. I wonder if homeschooling through high school would actually be that bad? Oh, Here comes my P.E. teacher. He's tossing us the ball and telling us to play. I don't know how to play basketball! I ask my friend; she doesn't know either. We're walking over to these girls on the court; maybe they'll help us. Uh, maybe not. I don't think they're interested. Okay. We'll just shoot toward the basket then.

This whole day has been like that. This morning, first period, a few people were giving our choir teacher a hard time, and one guy yelled at him. So the teacher made us sit silently for the rest of the class. Okay. Next period—geography. My teacher (who is also my P.E. teacher) took an unusual interest in the girls. Subtle, but noticeable. Frankie was persistently bugging me to let him copy my English homework. The only things I accomplished in this class were reading a chapter in the book, answering six questions, and writing a letter to my friend. Next period—physical science. I did like this teacher. He was loud and obnoxious. Fun to watch. But he's pro-evolution all the way. We watch two videos on volcanoes. Class dismissed. Next was lunch, and then that brings me back to P.E. Now I have two more classes to go and this day is over. Days like this are so monotonous when one knows she can do more with her life. I just want to go back home....

I remember that day like it was last week and not four years ago. I did decide to go back home—a decision I believe was God-led. Since then, I have accomplished so much without having to be restricted by the structure and boundaries of the public school system. When we started to homeschool again, my mom was thrilled. We custom picked our curriculum and hooked up with some homeschool groups.

Some of these groups were geared specifically toward teens, while others served homeschoolers of all ages. I have noticed that our society misleads us into believing that children must be separated into age groups to be able to learn. But, I could see that members of my family aren't all the same age as me, yet we are always learning from and with each other. I realized that God had placed people of all ages in my life purposely so that I could learn from them.

Because the church we had just started attending did not have any other teens, I was at first a little disappointed. But soon enough I learned that just because I was officially a "teen," I didn't have to restrict myself to friends my own age. I began to talk to the adults, and they became vital role models in my life. I found out that they were, indeed, interesting and fun to talk to, not so old and boring as I had thought before. I also made friends with the children at my church and, although I didn't know it then, I was a role model for them. I started teaching Sunday School and discovered the incredible impact I can have on children's lives.

Because homeschooling allowed me to be flexible, I did quite a lot of baby sitting and had a part time job at a music and art school helping with the children's performing arts program. I also had two younger sisters to help with their schooling. I learned to be self-motivated, diligent, and patient. As I look back on high school, I see that God, in His divine grace, has opened the doors for me to learn. I never had to go out and look, God just took me through each door.

However, at one point I feel I could have missed God's grace. There were a lot of spiritual challenges I had to face. Because I was homeschooled I was different from other high schoolers. When I told others about my schooling, I got one of two common responses: "Wow, it would be so cool to be home all day to eat, sleep, and watch TV," or "What about your social life? How do you make friends? I couldn't possibly make it through the day without all my friends around and only my mom." I got frustrated trying to explain that I didn't lounge around at home all day. On the days we weren't busy with field trips, park days, skate days, drama, P.E., or other workshop classes, I sat at the kitchen table or on my bed doing school work that was challenging, and, although it got tough at times, I enjoyed it!

People thought I was weird for what I believed. "Why doesn't RaeAnn party, or date, or dress like everyone else?" I came to a point in my life where I had to ask myself if I really believed what I said, or was it just because of my parents' rules? At that time, my walk with the Lord was shallow. So when I would talk with people—some friends and some others whom I met here and there—they would tell me about the parties they went to and the guys they dated. I began to envy their freedom. Freedom to come and go as they pleased. And their parents never seemed to mind. They would talk about what they did at school, and the activities they were involved in. And even though I'd already had a taste of public school and did not like it, I felt that maybe I didn't give myself a very good chance to experience it.

I slowly started conforming to the ways of the world, despite all I had been taught. I was lowering my standards, and before I knew it, I was missing out on God's awesome plan for my life.

Someone once told me that God knows we all fall short at times and that is why He gives us His Word to live by so that we can set our standards high. When we have low standards and fall short, we're really going to drop low; but when our standards are high and we fall short of God's plan, it is not as hard to pick up and rise again. So I began to see that what I was envying in my friends was counterfeit. I watched as they, one by one, fell. Whether it was drugs, drinking, or pregnancy, it all stemmed from being permitted to do whatever they wanted with no boundaries.

"Experience life and all that it involves," is what the media and high schools tell us today. I am grateful that I had the opportunity to be home and sheltered to some extent, not sheltered in the negative sense, but sheltered in an environment that was healthy. Just as a plant, when it's planted in good soil, with pruning, water, and sunshine will produce a beautiful, healthy plant, I felt that the shelter and nurture I received protected me so that I could become a "healthier" person. There is a spiritual law in this analogy. Without me even knowing it, through homeschool, God was preparing me for the job I have now. It is so awesome how He has our lives all planned out.

I had an idea of what I wanted to do when I graduated, but God had a different idea. In the beginning of my senior year I was offered a job at the House of Hope, which is a home for women and children. I would be an aide to the Children's Program Manager. I started working three months later, doing my school work at night and on my lunch breaks. I not only gained more experience working with children, but I got to see the consequences of the choices made by some of the women who were now living at the home to get back on their feet. That was a real eye-opener. Since I was only seventeen, younger than all of the women, it was difficult trying to gain respect from them. And because I was raised in a totally different environment from most of them, they had a lot of questions about my life, especially since I did not go to a high school—it was just so strange to them. Because of the pasts that these women have come from, dealing with the different attitudes and beliefs got stressful at times and I have wanted to give up. But I am thankful that I stuck it out through the times of friction because now I am reaping the benefits. I was recently promoted to being the Children's Program Supervisor. I love my job and realize that God had prepared me for it. The love I had for children and the desire to teach, God had put to use in my life. So switching from homeschool to this job wasn't such a challenge as it might have been. I'm eager to see where God will lead me next.

A lot of people ask about graduation. For me it was a little different than most. Because we weren't enrolled with an I.S.P. (Independent Study Program) we decided to have our own graduation with just me. I'm not the type of person who likes to be the center of attention, so when my mom came up with the idea, my response was adamant: "No thanks!" But she knew what she was doing and planned a graduation for me anyway. She invited friends and family to come, and it was held at our church building. My parents and my two sisters got up and shared a few words they had separately written. So did my best friend and a few other friends of our family. It turned out better than I imagined, and I'm glad I consented. But my mom did *not* get me to wear a cap and gown except for the picture.

In addition to my job, I also have hobbies that I pursue. Drawing, flute, and reading are my favorites. One thing that I have learned is to never give up on my hobbies. Because my job is full-time, I can easily allow myself to be consumed with work and responsibilities. Doing things that I enjoy just for the sake of enjoying them, not having to meet a quota, provides a diversion from the same routine every day. I also hope to eventually go on a short-term missions trip. For now, though, I am contented in my own city at my own little job. I am thankful to my parents for listening to God's voice when He called them to home teach me. For I have learned far more than what books can teach. I would encourage anyone who is homeschooling through the significant years of high school, or is about to start, to press on, learn all you can, and don't let yourself be ripped off by the world's view, because all that it offers is temporary. Most of all, delight yourself in the Lord and He will give you the desires of your heart.

Kathleen Melvin lives with her sister Adriel and the rest of her family in Memphis, Tennessee.

CHAPTER 7

I am the second of three sisters, and have had a long and varied schooling history. My dad was in the military for many years, so my family moved around a lot. I attended several different schools—private, public, parochial, and military public schools, but my parents chose to home educate my sisters and me halfway through my fifth-grade year.

My new homeschooling curriculum was not easy. We quickly saw that there were many basics I had not mastered, even though I had made straight A's in the other schools. I loved the work, though, because there was always a light at the end of the tunnel—as soon as I finished my assignments, I was done for the day, period. NO HOMEWORK!!! I was sick of homework by the ripe old age of ten. Not just tired of doing it, but tired of the terrible panic I felt every morning before school, as I stood in line out on the chilly sidewalk. I forgot my assignments often enough to keep me in a constant state of worry—my parents would help me remember and actually get me out the door with them if I got the assignments home or written down, but that was the real trick. I really enjoyed the freedom and lack of stress in home education.

I was known in school as "Motor-mouth Melvin," social director, confidante, and lender of every school supply
I owned. (My third grade teacher said that if she could
hook my mouth up to the Georgia Power Company, we'd
never have a power shortage.) I had a great time and lots
of friends. When I started homeschooling, though, my sisters became classmates, friends and playfellows. We still
had our school friends over often and were involved in
church and homeschool groups, but my family became my
daily focus, not strangers whose worldview and values
were unknown and often questionable. This has greatly affected my maturity, as I can see now that I look back. I
tend to be very dependent on what my friends think and
do, and homeschooling helped to wean me from this and
make me much more independent of my friends, yet much
more reliant on my family whose values I share.

I am now a college freshman at the University of
Memphis. My home education experience has held me in
good stead, and I am having a great time. Everyone asks
what the transition was like—if it was hard to sit in a lec-
ture after so many years of working on my own with a
book; if I had to adjust to the work; if changing to an institution with all the freedom of college was a shock after
years of quieter, more-controlled home study. I love to answer these questions—the only real change was carrying a
backpack!

Actually, home education at the high school level prepared me for college work in a lot of ways. The entire format of our homeschool was arranged so that I had a
schedule for my daily assignments and was responsible for
completing those assignments. If I had any questions, I
went to mom and dad for help. They oversaw my work
and kept me on target. This is so similar to what I've found

college study to be like. I was very responsible for my education in high school. I really believed that if I didn't learn, it would be my own fault and could greatly affect my future. I feel the same way at the University. The independent work and the amount of effort I put in is up to me and will show up on my GPA and transcript. Changing from home education atmosphere to a college or university was not a major shock to my system.

I've decided to double major in journalism and piano performance. It will mean a lot of practicing and hard work, but I want to try. I love music and could never have made it this far without the free time homeschooling allowed me. I started piano lessons when I was in third grade and continued into high school. When I was 11, I had the opportunity to take violin in a string class that our church was starting, but I didn't take lessons privately until I was 15. That is when we met an amazing Suzuki violin teacher, and I asked her to start me at the very beginning, as if I knew nothing. (Actually, I didn't know anything, and was it obvious! The man at the music store threw in three sets of earplugs when we bought my violin.) I had the time to practice up to two hours every day, got used to being closed up in the music room or playing to the rhythm of doors shutting, and progressed quickly. I'm in the tenth and last book now, hoping to graduate from the next two levels in May of 1997. I was in the Memphis Youth Symphony my senior year, and now I'm receiving more and more "freelance" performing jobs, which are a great financial help to this college student! There is absolutely no way that I could have earned the piano and violin scholarships that I did without the mental freedom, energy, and time I had during my home-educated high school years.

Many opportunities arise for me to share my testimony as a Bible-believing Christian. For instance, in a particular honors class that I was required to take in order to receive an honors certificate, I had to answer the essay question, "How do you arrive at your opinions, beliefs, and values?" Was that a blank check, or what? I had several Christian friends in that class, and we all received good grades. (I still haven't figured out what I am supposed to do with an honors certificate, though.)

I can gratefully see as I look back that I have been trained as more of a thinker than a learner through my home education experience. A learner is merely fed information, and accepts it much like a computer or a baby bird, without questioning the values of the hand providing it. A society of learners exhibits very few discoveries or innovations and creates the perfect environment for a dictatorial government. A thinker, on the other hand, searches things out for him or herself, carefully examining knowledge before it is received as truth. Think of all that would be different if Columbus had not challenged the theory of a flat earth, or if Pasteur had accepted the idea of spoiled meat producing maggots! In order to be commendable citizens and superb at whatever our occupations, we must truly think, discerning truth in every area, however trivial it may seem. This has been, perhaps, the greatest lesson of my high school education—a constant challenge in all aspects of life. And to think that I learned it as I sat at the feet of my wise and loving parents, alongside my sisters, my friends.

HOTHOUSE TRANSPLANTS

Danae Kershner is a practicing midwife who lives in Maryland.

CHAPTER 8

MAKINGS OF A MIDWIFE

Thus saith the Lord; cursed be the man that trusteth in man, and maketh flesh his arm, and whose heart departeth from the Lord. For he shall be like the heath in the desert, and shall not see when good cometh; but shall inhabit the parched places in the wilderness, in a salt land and not inhabited.

Blessed is the man that trusteth in the Lord, and whose hope the Lord is. For he shall be as a tree planted by the waters, and that spreadeth out her roots by the river, and shall not see when heat cometh, but her leaf shall be green; and shall not be careful in the year of drought, neither shall cease from yielding fruit (Jeremiah 17:5-8).

The sound of ringing beside my bed awakens me. The telephone insists on being answered even though it is 2:00 a.m. God's gift of adrenaline kicks in and I am immediately alert to Stewart's voice on the other end. He tells me that his wife Sarah is in labor and that contractions are five minutes apart. He thinks Sarah would like me to come now.

Quietly I slip out of bed. Shivering in the frigid air of my bedroom, I dress in my skirt and blouse, shoes and socks. With a last glance at my two sisters sleeping peacefully in their beds, I momentarily wish that I could join them for the rest of the night. But knowing that God has ordained this time and not wanting to miss His miracle, I turn and silently close the door behind me.

I awaken my parents who groggily ask where I'm going. My father reaches out to place a hand on mine and prays for God's blessing on me, on Sarah, Stewart, and the new baby. Then, armed with the prayers of the saints and my birth bag, I am released into the cold, starry night: quiet, peaceful, and alone.

Driving along I pray for Sarah. I pray for wisdom, understanding, and safety. I pray that the Lord would be honored, and I pray that the enemy would be bound. As is so common when preparing for a birth, I am flooded by the realization of my own insufficiency and then cast myself upon Him Who is always sufficient. My thoughts turn from Sarah to the past and how I ever found myself in this place to begin with.

As a young child growing up in rural Maine, I was a model student at the village school. Always ready to please and fearful of disappointing my parents and teachers, I found school stressful and demanding. I would experience stomach aches in the mornings due to nervous tension, and, although I excelled academically, I never liked school. My mother sensed my need and desired to keep me home. But she had no idea of how to begin and had reservations about home education's respectability. As a pastor's wife, she viewed our family as an example of the believers and hesitated at the thought of being radical.

And so it wasn't until my eighth-grade year, when homeschooling was becoming a more common practice, and my parents were feeling the financial pinch of private schooling and were not fully satisfied with its outcomes, that they began discussing the potential of bringing my brother and me home. My oldest brother had already graduated from high school, and my sister was not old enough for school yet. As strange as it may seem, although I did not like school, I was completely disheartened at the prospect of leaving and wept on my teacher's shoulder the last day of school.

The summer passed with uncertainty still in the air, and fall found me resigned to my destiny. But at the last minute an opportunity arose for me to return to school, and my parents took the offer. The Lord has an interesting way of accomplishing His purpose even when it entails giving us the desire of our hearts but sending leanness to our souls. That is the best description for my ninth-grade school year. Excessive stresses and pressures accompanied by the loss of a very dear friendship caused a deep humbling in my heart. On the last day of that year I was more than delighted to leave and eagerly looked forward to being home.

I am a firm believer that God has ordained families to be the primary classroom for life preparation and that He places in families the perfect blend of personalities, talents, skills, disabilities, and uniqueness to temper and mold each individual for His honor and purpose. When we came home to learn, our family relationships suddenly grew worse. Quarrels became more commonplace than before. This was very disheartening to me until I realized that we hadn't degenerated. We simply had never lived with each other all the time and therefore had not had to deal with yielding our rights and preferring one another above ourselves.

My mother and father had the most creative ideas by which to instruct us and draw us together in a spirit of unity. My dad designed a prayer and praise chart that was intended to match each individual up with another individual in the family. Probably the most outlandish activity we did was one directed by my mother. We had been studying blindness—physical and spiritual—and learning to see life as God sees it. To emphasize the point of instruction my mother had her four children blindfolded, then she requested that we prepare lunch with that handicap. We had to work together, around each other, and into each other as the case sometimes was. When it was finished and we removed our blindfolds to view the devastation from our lunch preparations, we were overcome with hysterical laughter. It was a lesson not soon forgotten.

The essence of life is not a job, achievement, or possession. It is relationships—the relationship of God to mankind, and men to one another. It was to this end of giving me a new love, understanding, and appreciation for those relationships in my own home, where I previously had a dislike, that the Lord pressed me into the uncomfortable closeness of home education. But it was also there that I began discovering the most valuable lessons which no diploma could ever reflect, no letters behind my name would ever identify, and which no audience would ever gather to applaud. They were the lessons that come too late for so many "successful" people of this world. Little by little as I responded by grace to the fiery trials, the precious jewels of forgiveness, gratefulness, sacrifice, faith, virtue, temperance, perseverance, godliness, brotherly kindness, and love began to be mine.

My parents saw weaknesses in my character, the most obvious being a withdrawal into myself while avoiding interaction with others. They requested that I participate in activities I would have preferred to forego such as public speaking, swimming lessons, and driver's education. But I learned to enjoy a challenge and meet an unpleasant task head on without procrastination. All of this became vitally important to my ministry as a midwife because midwifery centers around people and understanding how to best serve so many unique individuals.

While God was directing my steps, I was devising my way and constructing plans by which to expedite my future. Interestingly, I had never felt a deep need or desire for a high school diploma. The focus in our home had been more on life education than on homeschooling. My parents had always emphasized to us the need for a fierce loyalty to God's direction and a lack of peer dependence. Consequently, I never felt the need or saw the benefit of following the traditional course of high school graduation in order to pursue God's calling in my life. Many a well-meaning inquisitor has asked me how I would get a job without a diploma. One man asked this question immediately after he offered me a job as his employee! Ironically, the opposite seems to be true. I am never short of work! It might not pay $25.00 an hour, but I have never lacked for God's provision either. I find that the Lord's certification of character and integrity has supplied more work opportunities than my academic qualifications.

The Lord had given me a vision of not having a vocation but rather developing skills which would be used to meet an existing need. At the same time that these convictions were forming in my heart, my interest was leaning toward biology studies. I simply found anatomy and physiology fascinating! My mother, the wise woman that she is, saw this bent and suggested I consider midwifery training. Birth had always been close to my mother's heart, and she foresaw the void that could be filled by a Godly midwife.

Now midwifery was as new to me as I have found it to be so with many others. (One high school student said she had never heard of the term. Wasn't it when you get married but not all the way?) But as Scripture says in Genesis 24:27, "I being in the way, the Lord led me....," God began to open up doors of opportunity to see children born, to hear of the need and desire for midwives expressed, and to talk to those in the health profession who had worked with pregnancy and childbirth on a first-hand basis. Following all of that, I was convinced this ministry would not only provide a means of meeting physical needs but spiritual needs as well, and that it was definitely the Lord's direction for me. After humbling myself and seeking the Lord for confirmation, I was given the reassurance of Psalm 32:8, "I will instruct thee and teach thee in the way which thou shalt go: I will guide thee with mine eye."

But even in midwifery the Lord had unconventional plans for my training. It began with enrolling in an indepth home study course. After studying this for six months, I was offered an apprenticeship opportunity in Lancaster County, Pennsylvania. The three years I spent there were busy and full, giving me experience in the births of over 300 children. What an opportunity! And none of it was of my own manipulation. I am so grateful to my teachers who not only taught me about birth and academics, but about organization, loving families, life, wisdom, and truth through correction, instruction in righteousness, and life example.

At age 23 I returned home to be part of my family again and to minister to the childbearing families in the areas surrounding me. What precious events I have seen take place; what truths I have grappled with; what challenges

I've met; what places of peace I've found in the midst of storm and turmoil.

Perhaps the most significant lesson is that of faith. "Man devises his way but the Lord directs his steps" (Proverbs 16:9). Though I plan my goings, at times I am stripped of my ambition, my own power to achieve my ends and provide for my own needs. It is then, whether in the midst of a frustration at home or in a complicated birth where I am exhausted, that I cast myself, void and power-less, upon the mercies of the Shepherd and Bishop of my soul. "In the fear of the Lord is strong confidence; and His children shall have a place of refuge" (Proverbs 14:26).

A dog barks as I pull into the driveway. Stewart has been thoughtful enough to leave the porch light on. I take one last deep breath of the night air before stepping into the warmth of the dimly lit kitchen. I quietly let myself into the bedroom. I recognize the beautiful teamwork that these two have. Sarah glances up after her contraction is finished. I praise her with a word of encouragement and a gentle pat, then commence to set up and evaluate vital signs. The intensity in the room grows as depicted by the labored breathing and furrowed brow. I never cease to marvel at such human force as this. But such incredible ef-fort merits priceless reward, and in a short time I am gen-tly lifting a wet, pink, vivacious baby daughter into Sarah and Stewart's open arms. It's not long before mother and baby are settled clean, comfortable, and nursing into bed. We close with a word of prayer and thanksgiving, then with a last check on Sarah and baby Grace, I leave them settled for a well deserved rest.

Driving home I marvel at a beautiful sunrise and meditate on the goodness of the Lord in my life. To walk in the way of the Lord has been worth every sacrifice, every desire given up on the altar. Truly He is worthy of such honor. My mind turns to a verse I have recently read in Jeremiah 29:11-14a: "For I know the plans that I have for you, declares the Lord, plans for welfare and not for calamity to give you a future and a hope. Then you will call upon Me and come and pray to Me, and I will listen to you. And you will seek Me and find Me when you search for Me with all you heart, and I will be found by you...."

HOTHOUSE TRANSPLANTS

*Adriel Melvin lives with her
family in Memphis, Tennessee.*

CHAPTER9

I thought my parents had lost their minds when they decided to home-educate my sisters and me—surely they wouldn't do such a thing to us. It didn't take long for me to realize that it was probably the best step our family had ever taken. I was home educated for seven and a half years, and I am currently a sophomore accounting major attending the University of Memphis on a full scholarship.

Being home educated through junior high and high school was, as I said before, the best thing that could have happened to me. I grew and was strengthened in areas that only would have been stunted had I attended a regular high school. My family grew closer together and my sisters became my best friends—we were peer-dependent no longer. Instead, we came to be family-dependent. Before we began homeschooling, I was afraid to acknowledge my sisters in the school hallways. I preferred my friends to my family. That changed, though, as I came to appreciate the gifts and strengths of my family and was not influenced by my peers' opinions.

Homeschooling through high school was fun, and I enjoyed it more than any other time in my school experience. This is a profound statement considering the whole sampling of schools I have attended. As a "military brat," I attended eight different schools of all types—private, Christian, public, and military public schools—from California to Georgia, during my preschool to sixth grade years. I was always well-adjusted, loved my many teachers, made friends easily, and, oddly enough, enjoyed the newness of the different schools.

But I enjoyed homeschooling for different reasons; the foremost was that I enjoyed learning and discovering myself. It wasn't until I reached high school that I fully appreciated and utilized this aspect of home education. My parents gave me the books and curriculum I was to use as a guide and oversaw all that I did, but I was free to explore areas of interest. Home education freed me to work at my own pace—to take the time I needed to learn a concept, whether it was an hour or a week, before moving on. I also had the free time to do things I otherwise would have been unable to do, such as community service, working in music studios, handwork projects, and starting and maintaining two home businesses.

Homeschooling has prepared me for college, but also, and more importantly, for life. I was responsible for doing my schoolwork which ingrained in me good study habits and made me willing to work hard for high grades. Peer dependency became less of a problem as my parents taught me to stand for my values, alone if necessary, and not to be afraid of what others might think. My maturity also increased because I was the oldest in both my family and our homeschool support group—all the children looked up to me, and I was given responsibility for them. I learned teaching and leadership skills through our teenage girls group called "Kindred Spirits" where I was a leader and helped my mother teach skills to others. Through volunteer work I was taught how to relate to people in professional settings; during my high school years I volunteered over 300 hours to various businesses, museums, and libraries in Memphis. But the most important lesson of my high school career was learned through something as trivial as an art contest.

When I was in tenth grade, my mom told me that as part of my school work for that year I was to enter an art contest. I could not have been more surprised. Enter the art contest?!? I could not draw, paint, or sculpt anything that looked halfway recognizable, so why should I go through all the time, trouble, and humiliation for a high school credit when I could get it some other way? I told my parents that my sisters had all the talent in this area, and I was probably as unartistic a candidate as anyone could find. They stood firm and told me they just wanted me to do my best.

I fussed for a few days, but when I realized that there was no getting around it and that the deadline was getting ever closer, I sat down for some serious thinking. I came to these conclusions: 1) I can't draw or paint so all that remained for me to attempt was cutting; 2) One can only cut material, wood or paper, and I preferred to work with material; and 3) the art contest theme was the 100th anniversary of the ferris wheel and I enjoyed old-fashioned country scenes. As a result, I had a wonderful time making a cloth picture of a county fair, won the $100 "best of show" prize at the Mid-South Fair, and have since made two other pictures and have one in the planning stage.

I tell this story to illustrate the most valuable lesson I learned in high school—that I don't believe I would have learned as readily had my parents not home educated me. It sounds simple and very familiar, but I really learned it and will never forget it. I learned the value of obeying and respecting my parents' wishes in all things, not only when it makes sense or is easy to accomplish, for they see abilities, talents, and weaknesses in me that I have yet to discover. I also learned that the reward is much sweeter when a willing and submissive heart has been applied to the task.

When I graduated from high school, I was a little concerned about going to college. I was afraid that the work would be too difficult, that I wouldn't fit in on campus because I had been home educated, and that I would be generally miserable. I shouldn't have worried. I didn't realize at the time, but my home education approach to study had prepared me well for college. I was the one to arrange study groups and introduce people to each other in my classes. I was the one they came to with questions (little did I know that they all thought I was at least a junior), and all my professors knew me by name. The work itself was no harder than my senior year. The only difference was that I was required to write a lot more papers in college.

There were only two major adjustments I experienced when I entered college: the first was getting accustomed to carrying all my books around in a backpack (I was sore for several days); the second was much easier for me as I had anticipated its occurrence—the challenge of standing firm and not wavering in my Christian beliefs merely because I felt outnumbered. During my first three semesters at college I wrote several papers critiquing some very liberal literature and composing essays on subjects such as morality. In all of these I expressed very strongly my Christian viewpoint, so much so that I feared the grades I would receive from my liberal professors. In each case I made excellent grades and passed all the courses with A's. There have been some times during these semesters when I feared being too outspoken, but the Lord has rewarded my faithfulness and given me more courage for the next confrontation.

He has also surrounded and blessed me with many Christian friends. My sister and I are both active in the Baptist Student Union and the Reformed University Fellowship on campus and have been greatly encouraged just knowing that there are other Christians on campus striving to remain faithful to their God. In each class I take, I try to locate other Christians, and I have found at least one in nearly every class.

I hope to be a wife and a mother someday, but because my parents and I don't know the Lord's plan for my life, I am preparing both for marriage and singlehood. During my homeschooling years I learned homemaking skills from my mother. Now, while I am at the university, I am preparing myself for singlehood. This does not mean that I plan to launch a career and work as an accountant until I marry, but rather that I will be trained to successfully support myself and my sisters out of my home should anything happen to my parents.

When I think back over my homeschooling experience and the testimonies friends have shared, I recognize that fear has a very strong influence in decision-making for most home educators. I feared homeschooling in the first place because I thought it couldn't possibly work. Later I was afraid that my parents might be put in jail for home educating us, and during my senior year I was extremely fearful about college—wondering if I could make it without disgracing my family and other homeschoolers. All these fears are very real and plausible, but they should not be allowed to influence the decisions we make more so than the Lord's leading. Thankfully, the Lord blessed me with parents who were strong and faithful to the Lord's will, despite a hesitant, sometimes-protesting daughter.

I am very grateful for the academic education, spiritual lessons, and close family ties that I gained through home education.

HOTHOUSE TRANSPLANTS

*Paul Glader, 19, is attending the
University of South Dakota on a
journalism and political science
scholarship. He also works as a freelance
writer for magazines and newspapers. He
plans to continue working for newspapers
when he finishes school. His family resides
in Colton, South Dakota.*

84

CHAPTER 10

HOW GOD CHANGED A DREAMER TO A DOER

It was the philosopher Plato who said, "He who wishes to move the world, must first move himself."

Although the early philosophers taught a man-centered view of the world, some of their observations held truth. Plato's words reflect the need for basic initiative. Without that quality, a capable mind is reduced to frustration, a potentially skillful laborer becomes a lazy loafer, an aspiring young man or woman becomes a dreamer who fantasizes away his or her young years.

Initiative is like the small action of a surfer climbing atop his board when a big wave crashes toward him. God brings waves for everybody in life. Those who ride the waves are the ones who climb atop the surfboard and say, "Let's go."

The very act of homeschooling involves major initiative on the part of parents. For a young person to maximize the homeschool opportunity, both the parents and the young person must continue on the path of initiative they started.

INITIATIVE WANING

As a young child, I remember operating paper routes, lemonade stands, backyard circuses and sidewalk-shoveling businesses with my older sister. Our parents allowed us to take initiative in a lot of ways.

But as I moved into the preteen years, I sometimes doubted the value of homeschooling through high school. As my attitude declined, so did my initiative.

I can recall daydreaming for hours about becoming a great athlete someday, or of being wealthy or famous. Meanwhile, the algebra lesson open on my desk was on hold until I finished my dreaming. Somehow, it didn't connect with me that success only comes by more work and less dreaming.

I would fritter away time playing computer games, shooting basket after basket in the driveway, and tinkering on the piano rather than practicing hard and practicing right.

Every day I would pour over the sports pages of the newspaper, excising every obscure piece of sports trivia and news. This proved useful, I thought. For if ever I were engaged in a conversation with friends about major league baseball in the city of New York, I would be the only one to know such facts as the severity of Don Mattingly's pulled hamstring or Dwight Gooden's body weight, height, ERA, batting average, and name of his dog.

The bulk of my energy was devoted to the sport of wrestling. When I wasn't practicing or working out, I was replaying my matches over and over in the theater of my mind.

My parents recognized I lacked initiative in the things that matter most: Bible study, a love for God, and helping around the house. They saw that the initiative I did have came out on the wrestling mat. They knew that, ultimately, I needed to develop a vision for accomplishment that is profitable spiritually, financially, and mentally, rather than just physically.

Although they looked for worthwhile activities and encouraged me to exercise initiative, they knew that a person, at some point, has to discern for himself where his life is supposed to go and how he is supposed to get there.

Thankfully, that time came. I have to attribute a change in my life purpose to attending the Seminar in Basic Life Principles at age 15. I came away from it realizing that 90 percent of the things I did had no eternal value. I felt as if I had fallen asleep for a few years like Rip Van Winkle. Suddenly, I had an intense desire to catch up. I started by asking God to direct me toward more productive goals. I called my wrestling coach and told him I wasn't going to compete that year. I told my parents I wanted to be under their authority and wanted to find what God's plan was for my life.

We had been using the Advanced Training Institute homeschool curriculum for a few years. I started taking my schoolwork more seriously, paying better attention to my mom's writing lessons, and working fourteen hours a day, at times, because I truly began to love learning.

That year, I saw God honor my obedience.

Without dredging through details of exciting events that year, I tried to live according to the verse in Proverbs that says, "The precious possession of a man is diligence. The slothful does not roast what he took hunting."

A CHANGE IN DIRECTION

By the time I turned 16, I was writing a monthly newspaper column for high school students, apprenticing under an attorney, and my older sister and I were speaking regularly to church youth groups in our rural area of South Dakota on topics such as bitterness and self-acceptance.

Again, the ATI teaching on the value of learning different skills to determine life purpose was a big motivator at this time. And my parents encouraged my sister and I to begin looking for apprenticeships.

Because of the newspaper experience I had developed and a God-ordained contact at a large newspaper in Indianapolis, I was hired the next summer to work as an intern in the editorial department of *The Indianapolis News*. Originally, I was brought in to lay out the editorial page. But soon I found myself writing editorials. The three month internship turned into seven months. This first experience led to longer internships and part-time employment at that newspaper.

Through that job, God has shown me that the vocational part of my life purpose is to be a reformer in the mainstream press and to echo the Psalm where David prays, "May I publish with a voice of thanksgiving."

Over the past four years, I have learned many aspects of the newspaper/magazine field. In addition to work in Indianapolis as a reporter and editorial writer, I've spent time writing news, features, and sports stories for newspapers and magazines in Minnesota and South Dakota. Some of the publications I've written for include *The Rapid City Journal, The White Bear Press, The Minneapolis Star-Tribune, Group Practice Journal,* and *Minnesota Sports.* Additionally, I was involved with the start-up of two publications: *Young Conservative* magazine and *Pro-Family News.* Recently, I contributed a major section to a book titled *Indiana Legends,* authored by another feature writer at the *Indianapolis Star-News.*

One of the primary reasons I love the vocation God has given me is the variety. When you work as a reporter, you learn something new every day. In a way, it is a lot like homeschooling. I've learned how cities work by covering city-council meetings and city government. I've learned how the criminal justice system works by covering the courts and crime beat. With profile stories, I've learned the art of interviewing—of drawing someone's life from him and putting it on paper. I've learned a good deal about politics and writing style by writing editorials and columns.

Some Christians view politics as dirty and don't emphasize the Christian's role in civil government. Likewise, the media is also a whipping boy for people of faith. Regardless of whether the criticism is justified, more Christian's are needed in the high levels of these fields.

Although it is becoming more of a profession than a trade, the business of newspapers has traditionally been apprenticeship-friendly. It used to be that a lad could walk into the *New York Times* from off the street and work his way up from copy boy to star reporter. Many of the great American writers such as Mark Twain, Ernest Hemingway, and Meridith Nicholson started in this fashion.

I still believe there is room for young people with character, writing talent, and the will to learn to enter this field through smaller newspapers, especially with the help of a coach or mentor.

A CALLING FROM GOD

It has become clear to me that finding a vocational calling is the necessary and natural outgrowth of developing Godly character (at least it should be). Having a calling from God to accomplish a specific task in a specific field is like shifting a car from neutral into first gear. There must be a purpose for developing character. In my case, it is to glorify God through journalism.

The great Puritan Cotton Mather wrote a tremendous essay around 1701 titled, "A Christian at His Calling." In it, Mather describes the importance of finding God's purpose in one's vocation.

"It seems a man slothful in business is not a man serving the Lord. By slothfulness men bring upon themselves, what, but poverty, but misery, but all sorts of confusion," he wrote. "With diligence a man may do marvelous things. Young man, work hard while you are young: you'll reap the effects of it when you are old. Yea, how can you ordinarily enjoy any rest at night if you have not been well at work in the day?"

Mather continues, "Come, come, for shame, away to your business. Lay out your strength in it; put forth your skill for it; avoid all impertinent avocations. Laudable recreations may be used now and then; but, I beseech you, let those recreations be used for sauce, but not for meat."

Since the time God has shown me His calling on my life, I've tried to live by Mather's admonition, seeking never to have a spare moment in my day, but always trying to read and learn. Now, the danger for me is not a lack of initiative but making sure that the projects I'm involved in are the most beneficial to my calling and God's plan. And God continually has to remind me that skill and success only grow in direct proportion to the extent that I love and honor Him.

It has been helpful to understand that every area of my earthly life is part of God's kingdom. Such areas as church, family, civil government, jobs, and our individual lives are each a division of God's kingdom on earth. He wants us to place Him first in each of these areas, recognizing His dominion over those areas as the reason we are involved in them. This gives us an eternal perspective and causes us to seek to have dominion over the world instead of just being on earth for a ride.

In life, you often see two types of people who are comparable to the proverbial grasshopper and the ant, the sluggard and the diligent, and the fool and the wise man.

The first type sits around waiting for an opportunity to drop in his or her lap. The other makes opportunity happen by making phone calls, showing up at meetings, and taking on responsibilities.

The first type becomes jealous and sulky when he or she sees others who reach success. (The unambitious usually hate the ambitious.) The latter learns from people who are more successful, is inspired to higher achievement, and often ends up networking with the successful.

In the end, the first type of person receives the same condemnation as the steward in Christ's parable who buried the money his master entrusted to him. The other reaps the rewards of initiative. It doesn't matter how noble or successful he or she becomes. In God's eyes, that person is a success because he or she obeyed Scripture's teaching to "dwell in the land and cultivate faithfulness."

When the Apostle Paul wrote that "each one of us shall give account of himself to God" (Romans 14:12), I believe he was talking about what we do with our earthly life and how we follow our calling.

Our desire then should be "to number our days that we may apply our hearts unto wisdom" (Psalm 90:12). (Copyright 1997, Paul Glader. Used with permission.)

*Stephanie Hutchison attends
Biola University and lives with her
family in Westminster, California.*

CHAPTER 11

Have you ever been asked, "So do you go to school in your pajamas?" or "How do you socialize?" or how about this one—"Do you teach your younger brothers and sisters?" These questions, among others, are ones that I have heard my entire school career until I started attending college, two years ago. (The answers to the above are as follows: No, never; at church, or with other homeschoolers; and, not academically, no.) Of course, there's always the question, "Do you *like* being homeschooled?" which is hard to answer if, like me, you've never experienced anything else. Actually, I did enjoy being homeschooled, and I saw the value in it, for when I hit high school age, my parents gave me two options regarding my education: remain at home, under the tutelage of both my parents while also taking some group classes and junior college courses, or attend the local public high school.

Now, I must tell you that there are two things I very much enjoy doing. I have a passion for acting which means involvement in theater. The second activity I particularly enjoy is swimming, and I love the diving part most. The high school I would have attended had both a drama club, which puts on various productions, and a diving team as part of their swim team, which happens to be rather unusual.

Even though attending the public high school would allow me to participate in two things I loved, I chose to remain at home because I thought that the quality of education that I was getting was higher, and I was being taught from a Christian perspective with a Christian worldview. (As it worked out, I was able to get drama/theater experience both in homeschool drama groups and community theater, and I was on the swim team of a local Christian high school for a semester.)

I do not at all regret making that decision. On the contrary, I think that it fostered my love of learning. It also allowed my parents and me to accelerate my high school education so that I could graduate from high school in three years instead of four. This was an added bonus.

The education I received gave me a good foundation not only in Christian principles and lifestyle, but also in a Christian worldview, which I would define as a Christian philosophy, or reason for existence.

All of this leads in a roundabout way to where I'm at now. When it came time to choose a college, I chose a private, Christian, four-year university within commuting distance of my home. God worked it out in ways that are nothing short of miraculous for me to attend this expensive institution debt-free, although it requires much hard work and serious study.

I've always loved studying history, and started out in a major to become a history teacher. However, I decided over a long process which included teaching swimming all summer long, that teaching was not the ideal vocation for me. I think that God has gifted me in other areas. Yet, I still enjoy studying history and continue to do so since I feel it provides a valuable base for my other studies.

I mentioned before that I love to learn. I am fascinated with learning about ideas and how they impact and change culture and, further, how intellectual Christianity can impact and change our culture. I am especially interested in the political implications of cultural beliefs and the practical applications that result. For instance, Goals 2000 programs being instituted on the federal level have tremendous impact at the local level. Through Goals 2000, our federal, state, and local governments are taking on responsibilities and roles of raising our children in the interests of protecting them and raising their quality of life. Programs like these, and even the concept of federal government involvement in education, would never have come into being but for the historical impact of ideas. Obviously, this becomes a political issue when there are those of us who do not want the government to raise, or assist in the raising of, our children. There are many such political actions that dramatically impact the family, and they are the indirect result of ideas which have impacted people.

The way all of this determines which area I should major in I have not yet decided. What I want to learn incorporates philosophy, some political theory, and, of course, history. I *do* know what one of my majors will be, and that is communication studies. I want to be able to communicate effectively to others all that I have learned, and how it applies politically as well as otherwise. I simply trust and pray that God will continue to direct and show me where I can best use the gifts and abilities He has given me.

I think that had I not been homeschooled, I would be heading in a very different direction. I would not have the commitment to family, which is reflected in my concern over how the family is being impacted in today's culture and what can be done to change that. My parents have sacrificed a great deal to homeschool me and my younger siblings, and I only hope that I will live up to their expectations and the plan that the Lord has for my life.

HOTHOUSE TRANSPLANTS

Michael Thorpe will attend Loma Linda University and plans to become a physical therapist. He lives with his family in Garden Grove, California.

CHAPTER12

Hi! My name is michael and i were eddicated at home. My momy and daddy wanted to shelter me frome the reel world so they brung me away frome shcool to learn me themselfs. home shcool is nice. It is good too. Peepol says that i dident have enuff social Interackshun. i don't sea why. My parents say they will let me talk to sombody my own age when I is 21. Whoee! I can hardley wate. My ant thelma says they is teeching me at home so they can in doctranate me and fil my head with weird ideas[1].

Not a day goes by that I am not thankful that my parents took the time and trouble to "indoctrinate" me. My parents taught me not to lie, steal, or kill. They taught me to be a man of my word and person of integrity. They taught me how to respect authority and myself. If this is the result of indoctrination, I am quite content to be a victim of it. I don't claim to be perfect or even close to it, but I have a solid moral foundation to stand upon. The shifting ethics and relative morals currently taught in many "real" schools provide little foundation at all. I'm proud to say I

---

[1]   If you prefer to read a relatively serious version of this story, disregard the above paragraph and ignore all future footnotes. Thank you.

was taught at home, and I plan to give my own children the same advantage.[2]

The extent of my "real school" experience prior to college was one year of preschool followed by half a year of kindergarten. Looking back from the lofty height of my twenty years, I still remember being utterly terrified of being caught in "London Bridge is Falling Down." This insidious "game" involved herding a group of reluctant children through an arch formed by the hands of two fellow students, who would sing at you if you were unlucky enough to be in the arch when it fell. Facing the massive wave of solid embarrassment that followed was a scarring experience.[3]

Excepting the above terror of terrors, one of the biggest[4] problems of my "real schooling" seemed to be a lack of personal attention. I don't fault the teachers because there is no way they could be attuned to the needs of every child, but I remember them as mile-high giants who were

2   You are reading this note, therefore I assume you have relinquished your right to completely serious content. Please be prepared for periodical forays into silliness.

3   Footnote, n. A note of reference or explanation at the bottom which documents or supplements the text; a statement, action, or event that is subordinate to a larger or more significant one; an annoying literary device which causes one to lose one's place while attempting to read; a note written upon one's foot; a musical note created by tapping one's foot.

4   Some of my footnotes may appear to be totally out of context with the surrounding material, but I wish to assure you that the position of each has been fastidiously calculated to give the appearance of random placement.

oblivious of the little people below. I know they cared, but they weren't always aware of what was going on.

One example is indelibly stamped in my memory. A boy in my class knocked another boy down, then ran and told the teacher that the other child had done it to him. This happened several times but my teacher never had a clue as to what was going on. She ended up punishing the boy that was pushed down. He was a little "slow," and could not verbally defend himself, so he was made to sit outside. The sense of justice in a person peaks at five years old, so I went to the teacher and tried to tell her that she was mistaken, but it didn't even seem like she was listening. I kept repeating myself and eventually the boy was returned to the classroom, but the real bully went unpunished. This happened before I had heard the phrase "the world isn't fair" a million times, and it made a deep impression on me.

Then there were the times when I skipped entire sections of the "Dick and Jane"[5] type reading books without anyone noticing. I got bored with the lessons and just started flipping through books several pages at a time. My teachers congratulated me and said I was progressing rapidly. Little did they know!

About half way through kindergarten my parents decided to venture into the void and try this weird, new thing called homeschool.

Note: It will become necessary in the course of my writing to use the words "homeschool," "homeschooling," and "homeschoolers" with considerable frequency. In the interest of time, sanity, and to avoid writer's cramp, I have

---

[5] Classic Dick and Jane: See Jack run. See Jack sing. See Jack get eaten by a giant purple dinosaur.

taken the liberty of abbreviating these words as "H.S.,"
"H.S.ing," and "H.S.ers" respectively.[6]

The comfort and safety of home was the most intellec-
tually nourishing environment I could have wished for
during my early schooling. There was a certain freedom of
movement within my subjects not found in a traditional
school setting. I was able to follow tangents that interested
me or try to put to use what I had learned.[7] I also felt free
to ask as many questions as I liked, because they were an
accepted part of the lessons. One of the greatest aspects of
the H.S. environment, was the ability to take as much or as
little time as I needed. In more traditional school settings,
the class either advances at the speed of the slowest students in each subject, or these same children are dragged
along at a pace too fast for them. Because children all have
their own strengths and weaknesses, they are likely to be
going too fast half the time and too slow the other half.

Most people, and kids especially, tend to learn better
when they are interested in a subject and have some say in
how they study it. This style of learning promotes a general interest in school instead of the typical boredom and
apathy. Eventually, there came a time in my school career
when I needed specific lesson plans and team work, but
when I was young I needed the freedom my parents allowed me.

---

6   The previous note could not be included as a
footnote for the simple reason that by this time many
people have left off reading the footnotes entirely.

7   This account of my blissful and
intellectually-free childhood might vary slightly from
actual events. Hindsight may be 20-20, but it comes with
a complementary pair of rose-tinted glasses.

After my traumatic experiences at the hands of Dick and Jane,[8] I had a marked dislike for reading that lasted for some time. Then once upon a time, I read a book that interested me and didn't involve seeing Spot run. A whole new world opened up before me and I began to read for enjoyment! I credit H.S. and a household ban on daytime TV for my becoming a voracious reader, and I credit voracious reading for significantly extending my vocabulary.[9]

Our schedule was pretty simple. There was a list up on the fridge with the day's assignments, and when I finished them I was done for the day. I could take as many breaks as I wanted, but I couldn't do anything with my friends until I was done. It is needless to say that my school hours were less than regular. If I was feeling especially diligent on a given day, I could have my work done by ten in the morning. I could just as easily draw it out until dinnertime. This arrangement worked especially well if a person wanted to go to a certain "rodent-themed" amusement park (conveniently located quite close to where I live) on a Thursday when very few people would be there. The plan was as follows:

1.) Call homeschooling friends and coordinate with them to get all work for that particular day finished earlier in the week;

2.) Consult with said friends about the best way to pass off the trip as educational to the affected parental units;

3.) Reason with said parental units;

---

[8] Actually it was more of a "Dick and Jane meet phonics" book, and the names were changed to protect the guilty, but it had all the elements of seeing Spot run and "the tan ant can," etc.

[9] I also credit voracious reading for making me near sighted.

4.) Plead.[10]

In my early years, "park day" was one of the highlights of my homeschooling week. A whole group of us would head to a park, the mothers sitting around discussing lesson plans and curriculum(or so they told us), and all us kids taking off for the nearest playground. Sometimes the moms would come up with some sly plan for working in an educational lesson, but we would fend them off as best we could. If we had to learn the name of the tree we were climbing or the species of caterpillar we were holding,[11] it made little difference to us.

Park day was the early ancestor of our YMCA P.E. classes. The combined purchasing power of a good-sized group of H.S. parents was enough to hire a couple of YMCA-affiliated sadists who began to whip and beat us into shape.[12] The lesser, and thus more popular of the two taskmasters, allowed us to play games and get occasional drinks of water, but not before the first fiend was through with us. We should have known what we were in for when she told us she was an aerobics[13] teacher. All we wanted to do was head for the basketball court and start playing, but she ran us around the gym, made us do sit-ups and push-

[10]   What we came to know as the classic "educational trip" scheme was recently claimed by some parents as some sort of a motivational ploy. Don't believe a word of it!

[11]   The caterpillars were always popular with our mothers. Not!

[12]   And our parents had wondered why they got such a good deal.

[13]   Aerobic: requires oxygen. Aerobics: exercises based on the assumption that humans don't require oxygen.

ups, and, of course, turned on the sappy music and led us through what could only be described as aerobic torture. Though we griped and groaned, we soon began to appreciate the results and found that the ends justified the means.[14]

As I became older, group classes began to figure largely in my weekly schedule. When a subject exceeded a parent's level of expertise, a group class was called for, and a knowledgeable parent was called upon to teach it. Often a parent with a special skill, such as drafting or artistry would volunteer[15] to attempt to enlighten us. Our group normally consisted of between six and ten students, "enlightened" and otherwise, but despite fluctuations our core membership remained the same.

One of my favorite group class activities was our debates. There are few things more fun than getting dressed up and arguing with somebody in the name of education. We would write our "stirring" speeches and practice crushing the opposing arguments with devastating logic, only to forget almost everything in the heat of battle. Oh well.

Some might argue that "organized" classes are contrary to H.S.ing in its purest form, but there is nothing wrong with H.S.ers learning together as long as their parents are involved and aware of what is being taught. Though only one parent would teach the group class, some of the others would usually sit in and learn with us. This gave them the ability to monitor the teaching, curriculum, and their own children if need be.

---

14   My ends were sore for months, believe me!

15   Volunteering may not be the best way to describe these poor souls who, writhing and whimpering, were compelled to impart their knowledge to us.

At one time I was part of a larger H.S. group that also gave group classes but in more of a traditional school sort of way. For a long time I felt there was a difference in the two types of classes, but until lately I was unable to put my finger on it.

First of all, the classes were held in a classroom environment and not in a home. As simplistic as this sounds, there is a good deal of significance in the location of a group class. If parents are teaching in their own homes, they are on their own ground and command greater respect. In this "traditional" classroom there was one teacher, and other parents rarely sat in. This contributed to the "mob" mentality of the group, several members of which acted up and disrupted the class.

Another important factor was the class size and composition. In smaller groups, the parent who is teaching has better control and the kids involved can't hide in the crowd.

Also, in most of our group classes, our teachers had known us and, more importantly, our parents for a considerable amount of time. We knew if we goofed off in class our parents would hear about it, and they wouldn't necessarily side with us. In the larger "traditional" group, the teacher didn't know the parents as well, and there were too many of us to keep track of. Basically, the parents in this group were not closely involved, and the classes lost a good deal of the advantages associated with H.S.

One of the things I am most often asked by the uninitiated is how I handled the transition from H.S.ing into college. I have to admit, I wasn't exactly brimming with confidence. In the "old days" there was a limited supply of "those who had gone before," and we were sort of running blind. Though I was assured that my schooling would be adequate, I had a slight nagging feeling that I was inferior or ill prepared.

When I started at a junior college, I faced up to my fears and put what I had learned to the test. To my surprise and delight I was better prepared than the majority of the students around me. My first year was definitely a different experience than what I was accustomed to, as it would have been for any average high schooler, but I did okay.

Though I adapted quickly enough academically, I must say the social aspects of the college setting were slower coming. I had gone from a totally Christian-based environment into an almost completely secular one, and it took me a while to learn how to interact with and relate to my teachers and fellow students.

Though I am inclined to hold up H.S. as the perfect answer to the education question, there are those who would point out disadvantages. For instance, when I was younger, the first thing out of a peer's mouth when they found out where I went to school was, "What about the *"Prom"*?!!*[16] Well what about it? If attending a *Prom* is the definitive measure of a person's social interaction or success through high school, I would have to say that I was sorely deprived. In fact, whole groups of us H.S.ers would assemble and be socially deprived together. While limited social interaction might be a legitimate charge against homeschooling in some few cases and in a very few places, it becomes less so every year. The ranks of H.S.ers are increasing exponentially, and the possibilities for fellowship are ever widening.

---

[16]   Prom, n. A college or high school formal dance; the social climax of early life, without which one is deemed deprived.

The one thing that I felt I missed as a H.S.er was school sports. I always wanted to play football, and I almost joined a "real" school to do so. This is no longer as much of a problem considering that some private schools and specially organized groups have begun accepting H.S.ers to their teams.

Another objection I have heard from those uninitiated in the joys of home learning is the fact that H.S.ers are sheltering their children from the real world. Apparently, a person just hasn't lived until they've witnessed their first gang fight. One only has to read the paper to discover that life isn't "G-rated," and, frankly, I'd prefer my kids[17] to be sheltered from it than to have them drown in it. Wouldn't it be more damaging to expose a child to too much violence and hate than to hide it from them? As it is, we H.S.ers are not completely blind to the world around us. We are simply able to watch the world from the comfort of our safe haven.

To end on a positive note, the more I talk to people about H.S., the more accepting they seem to be. In my early years no one had even heard of it, and I spent a lot of time explaining the basic concept. Nowadays, I have found that the majority of people know what it is, and a good number know someone who is doing it.

Take heart! We are part of a growing and dynamic movement and have no reason to be either afraid or ashamed of what we are doing.

---

[17]   My future children, of course.

HOTHOUSE TRANSPLANTS

*Samuel K. Sanseri lives with
his family in Oregon.*

CHAPTER 13

SECONDARY EDUCATION: A FALSE DILEMMA EXPOSED BY THE
REAL-LIFE EXPERIENCES OF SAMUEL K. SANGER

In considering home education as a viable alternative
to private and government schooling, a false dilemma often emerges. Homeschoolers face the supposed tradeoff
between academic excellence and spiritual preparation.
Which is better, a smart, sociable, heathen student, or a
stupid, antisocial, spiritual one? Since the spiritual advantages of homeschooling are generally recognized, I will
use personal experience to refute four myths on the
"worldly" side of educational preparation.

MYTH #1: IF I HOMESCHOOL, I WON'T BE ABLE TO PLAY HIGH
SCHOOL SPORTS.

The burning desire to play basketball haunted my
mind during my senior year, though I had little hope of ful-
filling it as a homeschooler. I had played on community
teams in the past, but never in high school. Then I remem-
bered what the Boys and Girls Club director told me after I
played in eighth grade: "We don't organize high school
teams, but if you provide the coach and players, we will
supply the referees and competition."

"Do I know enough high school homeschoolers to make a team?" I wondered. "Who will coach us?" The questions raced through my head. Our homeschool support group had a list of families, including the names and ages of the children for each family. Our homeschool teen support group, the Home Education Association of Teens for Christ (HEAT) had a similar list.

So I began calling the high school guys about the possibility of starting a team. Almost all of them expressed their interest, unleashing their questions: "Who will we play against? How many games will we have? How many practices per week? Will there be tryouts? How will we get transportation? Where will we practice? Where will our games be?" I told them to check with their parents and I would call them back.

My initial ambitions exceeded logistical reality for all of us. I wanted five practices per week, starting in the summer. I planned for our team to challenge some of the local schools to a game. During my conversations with the other guys, I realized my expectations would have to be toned down a bit. Most of the guys could only handle a commitment of two nights per week for games or practices. This eliminated the possibility of challenging school teams since most school teams practice every day.

After the first round of calls, I thought I had enough players to make a team, so I decided to pursue a coach. I knew Mr. Hall would be just the man. In scrimmages at our HEAT meetings, I had seen him sink difficult shots, time and time again. He directed his whole team's ball movement to devastate the opposition. Later, I learned he had played in college and in one game scored 26 points. I knew his work schedule kept him extremely busy, but I decided to ask him anyway. When I called, Mr. Hall asked many of the same questions as the guys. In the end, he agreed to coach us! My joy knew no bounds.

When I phoned the guys to let them know we had a coach, however, my joy turned to grief. "I can't play," said one. "My parents are opening a restaurant and they need me to work nights." Another had medical problems and wouldn't be able to play. A third lamented that we wouldn't be able to practice five days a week or compete with school teams. Still another responded that he didn't have transportation and even if he did, he didn't want to play after all. He preferred baseball. "I don't know if I'll ever get a team together," I worried.

A few days later, a homeschool dad new to the area called, expressing concern about his son's need to participate in sports. When I told him about my plans, I could sense his excitement. His talented son would join the team, and he volunteered to coach! Mr. Hall appreciated having an assistant. In fact, his work required so much energy that he asked Mr. Friesen to be head coach and allow him to assist. The Friesens also knew a couple of other interested homeschool guys. Before I knew it, we had a full team and not one, but four coaches!

Our season was almost anticlimactic. In spite of our modest 4 - 5 record, we enjoyed playing together! Our supporters outnumbered the opposing fans by about three to one. The team continued on for several years without me, improving each year.

By God's grace, our basketball team debunked the myth that high school homeschoolers can't play sports. We balanced basketball with our other priorities and we played on a God-fearing team with God-fearing coaches and fans. Not only did I play basketball, but I gained satisfaction from initiating something which became a blessing to other homeschool families.

MYTH #2: IF I HOMESCHOOL, I WILL BE LESS PREPARED FOR COLLEGE.

Scene 1: The Placement Exam

I hadn't been in a classroom since the fourth grade and I was nervous. College hopefuls filled the examination room at Clackamas Community College. As the aide distributed the booklets, she asked if anyone wanted to take the advanced math test. I certainly didn't want advanced math. I only wanted the beginning college math test so I could take college algebra. No one else in the room requested the advanced test, so I didn't either. When I read the front cover of my test booklet, however, I realized I needed to take the advanced test in order to qualify for college algebra. I studied the sample problem on it for a minute, then for two minutes. It looked hard. Almost ready to give up, I looked at the problem once more. It differed from the algebra problems I had done in high school. Then I recognized it as a comparison problem, where two quantities are given and the student must determine whether the first is greater than, less than, or equal to the second. "Oh, this is simple," I thought. "Okay, I'll take this exam," I told the proctor.

The math test went quickly. An English test followed. I waited along with the other students while an instructor graded the exams. Finally, a smiling lady called my name. "You've placed into college writing and reading," she beamed. "You can also take calculus!"

Scene 2: The Trigonometry Final

I decided to delay calculus instruction until I learned more foundational math. At sixteen, I knew I had plenty of time to take college algebra and trigonometry first. After I passed college algebra, the trigonometry class zoomed by. I did my homework in the slow, methodical way I had learned in high school: do one problem at a time; write neatly; when struggling with a concept, review it until I understand it. I didn't realize how well I had done in the class until the professor told us, "Make sure you study hard for the final. All of you need to do well on it. There is only one person in this room who would pass without taking the final." He pointed at me.

Scene 3: English Composition

"Shivering in the middle of the stream, I dumped the bucket of water over my head, shocked out of my drowsiness by its iciness," my masterpiece began. My first real college paper, it described my summer trip to Honduras to work with missionary friends. I could hardly wait from the time I submitted it until the professor would return it with the beautiful A on the top of the page, perhaps with a few good words like "Excellent" or "Magnificent" thrown in.

When I received my paper back, however, it bled with corrections. Discouraged, I took the paper home and examined the corrections more closely. I disagreed with the professor on some of the mistakes he had marked. In particular, he marked as a spelling error a word that I knew I had written correctly, so I found him before class.

"I thought I had written a masterpiece," I told him, "but I'm discouraged by all these errors you marked." When I showed him his mistake with the spelling error, however, he apologized, and urged me not to give up, explaining that he thought I would make a good, descriptive writer and that I only needed to polish my punctuation. Heartened by his remarks, and determined not to repeat my mistakes, I went on to earn an A in the class.

In community college, I struggled through every class. Sometimes I got down to the wire before an exam, realizing I would not be able to read all the required material. I agonized over selecting the most important information and trying to memorize it. Often I could not sleep because of the tension before a deadline or exam. The members of my church's prayer meeting will tell you how I agonized over each project and test! However, the training as a homeschooler helped me prepare for exams, write research papers, and study independently. I received high grades, joined the Phi Theta Kappa Honor Society, was selected for the National Dean's List, and was admitted as a transfer student to Portland State University.

MYTH #3: IF I HOMESCHOOL, I WILL BE LESS QUALIFIED FOR SCHOLARSHIPS.

The instructions for the essay portion of my application for the sizable Oregon Laurels Scholarship at the university stated, "Discuss below in a concise, typed, statement, your personal, academic, and career goals. Inclusion of autobiographical information will be helpful."

Two opportunities during high school gave me an enormous advantage over my traditionally-educated competitors. During my junior year, Mrs. Ball, a Christian school administrator with homeschool connections, heard of my math aptitude and requested an hour per week tutoring session for their three freshman algebra students. My senior year, Mrs. Ball hired me to teach three levels of algebra to nine high school students for six hours per week, grading homework and exams. The unusual teaching experience and Mrs. Ball's strong letter of recommendation improved my chances of winning the scholarship.

A computer programming apprenticeship also came my way through homeschooling. Mr. Forster, a homeschool father, knew my family, and encouraged me to try my hand at computer programming under his tutelage. At first, playing computer games interested me more than bit-twiddling, but my parents underscored the value of the job training the apprenticeship provided, and I grew to enjoy it. Along with my "schoolmaster" experience, the apprenticeship practically put the scholarship in my lap. I only needed to write a good application essay.

I knew that because of the high value of the award, the scholarship selection committee would receive a large number of applications. By the time they had read mine, they would already be weary of reading about students' grade point averages and stereotypical career plans. I decided to write my essay as a story about my unique experiences as a homeschooler. Apparently, that topic caught their attention, and they awarded me the scholarship.

MYTH #4: IF I HOMESCHOOL, I WILL HAVE DIFFICULTY FINDING CAREER EMPLOYMENT

"I got the job!" my friend exclaimed as I entered the classroom early one February morning. Tom and I shared three classes as Computer Science majors at Portland State University. His friend from church had told him about an opening at Intel Corporation, the world's largest computer chip manufacturer. Tom applied and went through a month-long interview process before Intel hired him. I rejoiced with Tom as he told me about all the advantages of his position at Intel. I had been praying for him since he had told me of the opening a month before.

117

Work experience before college graduation is one of the most important preparations for a career in the computer industry and now Tom would have that experience with a company that any computer firm would instantly recognize. In the computer industry, a college degree means an automatic leap in the pay scale, so companies prefer experienced graduates over those who are inexperienced. A homeschool father who is a senior software engineer at Analogy, another large computer firm, exhorted me that I should begin actively searching for an internship to get that all-important experience, so I knew Tom's fortune in securing the position. I was a little jealous, too.

For the next month, I searched for any sort of computer job. I wrote my resumé using a lot of active verbs and sent it to my software engineering friend at Analogy. I read want ads in the newspaper. I scanned bulletin board postings in the Computer Science department at school. Everything required far more experience than I had. Then, one day in March, Tom told me, "There's an opening in our department. Here's my boss Jeff's business card. Just send him your resume and tell him you're interested in the position."

I sent my resumé with a cover letter to Jeff, explaining what my friend Tom had told me about his pleasant experience at Intel and about the opening. Then I prayed. My parents prayed. My church prayed. God answered.

I got a message on the answering machine. "This is Jeff from Intel. I just got your resumé and wondered when we could schedule an interview. Please call me back." He sounded eager!

The next morning I left a message on Jeff's answering machine. When he hadn't called back after two days, I called him again. This time, he answered. "I've passed your resumé on to Sara," he told me. "The opening is in her department." He gave me her phone number, and I called her. She didn't answer her phone, so I left a message. After a few days, when she didn't call me back, I called her. This time, she answered. "I've passed your resumé on to Kirk," she said and gave me his phone number.

I had grown quite discouraged by this time. At first, they were going to give me an interview, and now it seemed that they were trying to weed me out by passing my resumé on to other people. To make matters worse, Kirk would perform the interview! Tom had warned me about the difficult questions Kirk had asked him for his interview.

I called Kirk, expecting another answering machine. He answered the phone on the first ring. "Intel, Kirk speaking."

"Hi, Kirk. My name is Sam Sanseri. I am interested in the position in your department. My friend Tom told me about it. Sara told me to call you."

"Thanks for calling," Kirk said. "I called to find out when you would be free for a phone interview. Since I have you on the line, could you do it now?"

A phone interview! My mind whirled. I had been expecting an appointment for an on-site interview so I could study the things Tom had told me were important for the job. "I guess I'm free," I gulped. "I mean, I really wanted to study about interrupts before the interview."

"That's okay," Kirk said. I could almost hear the glee in his voice. "Now I'll get to see what you really know."

He proceeded to barrage me with technical questions: What significance do the words, "code, data, and stack" have? What is a semaphore? How do interrupts work?

How long is the biggest program I had written? I only knew the answers to about half, and even for those my answers were faltering. Then he asked me the fateful question, "Have you ever taken the cover off your computer?"

"He *would* ask that," I thought. I had only taken the cover off my computer once before and had broken my hard drive! "Yes. Only once," I explained. "I tried to install OS/2 2.1 and it required me to install it in my A: drive, but the disks I had were 3.5 inch diskettes, and my 3.5 inch drive was my B: drive. Someone told me I could reverse the order of my floppy cables and fool the computer into thinking that the 3.5 inch drive was my A: drive. Somehow in the process, I managed to break a pin on my hard drive, a $300 repair."

The interview continued as before, and finally, it ended, leaving me in the depths of despair. I could imagine their reasoning, "If he's this careless with his own equipment, how careless will he be with ours?" Now I had to do the most difficult thing of all, wait. Kirk had told me, "We'll call you. We still have more people to interview. It'll be about a week."

Discouraged about the interview, I called my friend at Analogy back, asking if he might have a job for me. Another friend of mine had also applied for the opening at Intel and had gone through the same interview process. After a week, he got a call saying, "We can't use you right now, but we'll probably want you about this time next year." Polite rejection. When I got home from school the same day, a message on my answering machine awaited. "This is Sara from Intel. I'm calling about the position you talked to Kirk about."

"What does this mean?" I wondered. "Will I get to finally come in for a real interview or are they going to give me a polite rejection like they gave my friend?" I called Sara the next morning.

She blew me away be asking when I could start and how many hours I could work during the school year. I replied that I would prefer beginning after spring term ended and that I could work Tuesdays and Thursdays during the year. She asked me to think about possibly working Saturdays as well and promised to call me back to make me an offer.

Thirty minutes after I hung up, my friend from Analogy called. "Hi, Samuel. I just got your message that you left a week ago. I haven't checked my voice mail for a while. I would really like to give you a job. I will have to convince my boss, though." Wow! Two job offers in one day! I told him I had just gotten a job offer at Intel. He asked me what I thought they would pay me and requested that I wait until hearing from him before accepting the Intel position.

I hung up and told my parents what had happened. My mom suggested that I get the advice of another computer professional whom we know through homeschooling. I called him, and he offered me a job, too! Three job offers in one day! Which one should I accept? After praying about it, I had it narrowed down to the first two job offers. The third had made a comment that I would not be worth what they were paying me, but they hoped that I would stick around long enough to make me worth their while. I knew I couldn't live like that. After more prayer, my friend from Analogy called back and regretted that he couldn't hire me because his boss had found someone with more experience for the position. Then I received a call from Intel with the offer, much more than I expected!

My life experience which began during high school homeschooling years made me desirable to the company. My resumé clearly reflected my testimony as well. This did not hurt me. I later learned that two of three of those involved in the hiring are Christians!

Getting the job did not complete my career goals. My position is in testing, often considered the low end of the totem pole by software engineers. I intend to be a software engineer by the time I graduate next year. When I began work at Intel, I noticed that our department had no documentation. Using the writing and computer skills I learned in high school, I wrote a fifty-page training manual for our testing procedures. The manual has already earned me enormous respect. I might be promoted sooner than I think!

THE SPIRITUAL ADVANTAGE

This story would not be complete without a small description of the spiritual benefits conferred on me through homeschooling. The abundance of time dramatically affected my spiritual life. I had time to establish the habit of reading the Bible every day. I committed huge portions of Scripture to memory by writing verses on index cards and reciting them to myself. I would not have had time to do all this if I had been attending school eight hours a day and coming home to do homework. Now that I'm in college, I don't have much time to memorize scripture, though the habit of Bible reading has withstood the time pressures!

My final homeschooling years, as I grew from boy to man, from student to teacher, and from apprentice to trained employee, formed the capstone of my academic and spiritual preparation for the real world. Since graduating, I have regretted that I did not form even more good habits during high school, since time shortage makes it more difficult these days. Homeschooling allows much more time for habit-building than do traditional learning modes.

In addition to the time factor, the worldview aspect weighed heavily in my favor. In a public school, the wisdom of this world prevails, and even in a private school, doctrinal emphasis might conflict with personal biblical understanding. On the contrary, my spiritual beliefs were nurtured in my home studies. After graduation, I spent a summer with missionaries in Honduras to discern God's will for my part in the Great Commission. God has not revealed His plan to me yet, but I must prepare myself to hear His call when it comes. It is not always a specific call like God gave to Samuel, my namesake. Sometimes it is a "whom shall I send" call as He made to Isaiah. I must take heed to my life and to biblical doctrine, study His Word, and pray, waiting on the Lord.

Another spiritual concern remains to me: finding a wife from the Lord. Homeschooling minimized the distraction of daily contact with girls my age. Some readers may wonder, "How will I meet my future spouse if I'm cooped up in the house all day?" I would remind them that it is only necessary to meet that ONE special person that God has prepared to be their lifelong partner. Through homeschooling, I have met quite a few godly young ladies. In His time, God will reveal His plan for me. I will wait until my "fields are prepared" (college completed and career begun) before "building my house" as God exhorts in Proverbs. As a senior in college, my debt-free status and promising career prospects accelerate my readiness for marriage.

This is an exciting and dangerous time in my life. I must spend much time with the Lord to overcome temptations and remain pure. I trust that my parents will give me insights that I might otherwise miss, and I continue to depend heavily on their counsel.

Home education has helped me in every way imaginable—in athletics, higher education, career pursuit, and even in my spiritual development. Homeschooling is the choice for those who prefer the dual edification of character and knowledge. The fear of the Lord is the beginning of knowledge and all treasures of wisdom and knowledge are hidden in Christ. Is it any wonder that academic excellence goes hand-in-hand with spiritual readiness? There is no dilemma!

HOTHOUSE TRANSPLANTS

*Belinda Scarlata lives with
her family in Madison, Tennessee.
She runs Family Christian Press.*

CHAPTER 14

MY HOMESCHOOL ADVENTURE

My homeschool adventure began when I was a teenager. My parents homeschooled my brother for about a month. Shortly after, they decided it would be best for all four children to be homeschooled. I was informed of the decision after I had made the cheerleading team. Needless to say, I was not happy. We were in a small, Christian school and came home to our newly converted family room/homeschool room, which included five desks, some bookshelves, a chalkboard, maps, and an eight-inch American flag. Thrills! The student body consisted of my two sisters, my brother, a toddler, and me.

I resented being taken away from friends, basketball, and cheerleading; but as time passed, and I was away from the worldly atmosphere, my outlook changed. Mom made it a priority to incorporate basic biblical principles and character quality training into our daily academic lessons. God started changing my vision. I began to see my old schoolmates through different eyes. Suddenly, the girls seemed giggly and immature. Their pranks, which I used to find funny, now seemed foolish. As the years passed, I began to notice that a lot of things about myself were changing. My priorities, goals, interests, and everything else were now defined differently than before.

More important than any academic skills, homeschooling has afforded my parents the opportunity to give me a solid biblical foundation in God's Word. Life's greatest lesson for me has been dependence on God. He is the answer to all things. When I totally trust in the Lord, no matter how a situation looks, He always directs my ways. His perfect will is what I want for my life. The Lord has mercifully given everyone an instruction book that contains his perfect plan. If we will only heed His Word and let Him have His way with our lives, then we will know that our lives will truly be fulfilled. The Bible has been my teacher, and guides me through important and everyday decisions on how to handle business, relationships, money, courtship, marriage, and many other things.

When I learned to trust God, I learned to trust my parents and their decision to homeschool. At first, homeschooling was such a change to my normal routine and surroundings that I rejected it with everything in me. But, like anything else, it takes a while to form new habits. With time, I became accustomed to the idea and accepted the fact that homeschooling was going to be a way of life for my family and me. One thing that I really admire about my parents is their steadfast will in their decision to homeschool. Many times I can remember going to my parents' friends and feeding off their sympathy. They would, in turn, encourage my parents to let me go back to high school. But no matter what anyone said, their decision was firm. For them, homeschooling was a commitment—just like marriage. Divorce was not acceptable, and neither was letting the children go back to school.

The time soon came when I began to enjoy homeschooling and our new way of life. Big changes occurred in all areas of academics. My mother had encouraged me to read while I was in school, but I found it boring and a waste of time. While homeschooling, I began to really enjoy learning and became an avid reader. In our home, everything became a learning experience. We couldn't even take a bath without learning about the economical savings a shower would render. I was actually finding this interesting.

So what do you do when you are happy with all that life offers? You ask for another challenge. I desperately wanted a job, but my parents objected to allowing me to work in an ungodly environment. Consequently, my mom came up with this bright idea: "Let's start a home business."

My sisters and I began small by selling homeschooling books to support groups. We were eventually traveling to curriculum fairs. During these activities, we were learning buying, selling, record keeping, geography, and basic computer skills. After a while, mom and dad felt the need for a permanent place where the homeschooling community could shop, so we opened a small homeschool store. Suddenly, we were getting practical lessons and training in customer service, profit margins, and accounting. Having a homeschool store opened my dad's eyes to another need—an umbrella school. Family Christian Academy was founded in 1990 to meet that need by providing support and legal covering to home educators. We began accepting enrollments and maintaining cumulative records for homeschooled children (while we learned how to file!). Soon, we developed a homeschool catalog because the selection of books was growing so quickly. During this time, we all got experience in desktop publishing, wading through postal regulations, packaging, shipping, and other practical skills.

As we became more involved in homeschooling, we became more aware of homeschoolers' needs. When we saw a need arise, we tried to accommodate homeschoolers by expeditiously addressing the need. It was really neat how each member of the family became a participant in the family business. My dad was the business manager and did all the purchasing, accounts payable, and lobbying for homeschool rights at our state capitol. Mom enjoyed writing, so she wrote for our catalog and newsletter. She has also used her homeschool experience to author a number of books to help home educators. I enjoyed talking with homeschooling mothers, so it seemed natural for me to manage the bookstore. One of my sisters found comfort in working behind the scenes, so she managed our student records department. Another sister, who loved the phone, took orders and helped in the shipping department. My brother and baby sister put labels on newsletters and handed out catalogs at curriculum fairs. Our family was later blessed with two more children for a grand total of seven. (I told my parents there were easier ways to get help, but they reminded me that no employee would work just to go to McDonald's once a week!)

Weeks turned into months and months into years. Before I knew it, my time as a student neared its end. In 1992, I graduated with six other seniors in our local support group. I can now look back, five years later, and say that I am just beginning to really learn about life and the Lord. Every day is a wonderful new learning adventure.

In 1996, I started my own business called Family Christian Press. My parents have played a very supportive role in my business journey. I've always sought their wisdom and I take advantage of the fact that my dad's office is one step away if I need an advisor. Family Christian Press is a distribution center for Christian bookstores and homeschool businesses. We provide a one-stop shop for businesses throughout the country who want to offer homeschool products to their customers. We currently have eight full-time people on staff and are continuing to grow. Nowadays, my time is spent developing and organizing new programs to implement in our business. The newly-added homeschool consultant program has taken off very well. The program provides homeschooling mothers an opportunity to train their children in business and earn money at the same time. The Lord has been good to me and has blessed our business efforts. I will be eternally grateful for the academic instruction, spiritual guidance, and apprenticeship opportunities my parents have given me. Finally, I know that homeschooling is God's perfect plan for my family.

My hopes and dreams for my future involve a husband, many children, a new career move called "homeschooling mom," as well as continuing in the call the Lord has given me with Family Christian Press, while I continue seeking Him for further direction.

(Copyright 1997 Family Christian Press. Used with permission.)

Christina Grace Thornton is the 21-year-old daughter of Claiborne and Lana Thornton. She graduated from homeschool in 1994. The events related here attracted national attention and were described in an article on the Eagle Forum website titled "Tennessee Slows Down School-to-Work: A Case Study in Successful Grassroots Action."

132

CHAPTER15

MY PART IN THE BATTLE AGAINST SCHOOL-TO-WORK IN TENNESSEE

When my parents started homeschooling my brother John and me fifteen years ago, people thought they were crazy. But while I was growing up, my parents protected me from the disapproval of their peers, and they surrounded us with people who thought much like us. They were not homeschooling us for social approval or because they thought that John and I would become extremely smart. They homeschooled us because the Lord gives the responsibility to parents for training and educating their own children. The first homeschooling family we met was through the newspaper; we discovered them when they were arrested for truancy. When I graduated in 1994, I was one of the first in Tennessee to have homeschooled for all my schooling.

It is not easy being on the crest of a new wave. As a Christian, though, I know that I serve a sovereign God who is faithful to put His people where He wants them when He wants them there, and also that He works all things together for His glory and for the eternal good of His children. He is always faithful to provide for all of our needs, including the strength to stand alone if that is what

He has called us to do. So, as I see it, when the Lord puts you on the crest of a wave, you thank Him and surf.

In the fall of 1995, I had just completed a number of weeks of service to a dear family homeschooling in the midst of a medical crisis in northern Virginia. The Board of Directors of the Tennessee Home Education Association (THEA) requested that my brother John and I serve them in the capacity of Legislative Interns for the 1996 legislative session. Our legislative session begins in January and concludes in the late spring. A full General Assembly lasts two years. This year and last I registered as a lobbyist for THEA and worked as an intern volunteer for THEA at the Capitol in Nashville. Mrs. Bobbie Patray, the Eagle Forum lobbyist who has lobbied in Tennessee for the past 20 years, has been a mentor to me.

This year our greatest battle is fighting School-to-Work (STW), or as it is called in Tennessee, School-to-Career (STC). Tennessee applied for a federal grant to implement STW/STC, and their application was accepted in August of 1996. Mrs. Patray got a copy of the Implementation Grant Application by "accident." She had called for some other information and the person said that they would send her a copy of the grant they had just completed right away. After Mrs. Patray read the grant application, she sounded the alarm to the other state pro-family organizations. The pro-family organizations formed a coalition to fight STW/STC—this coalition goes by the name of the Tennessee Education Information Project. The coalition began holding meetings to develop strategies on how to stop STC in Tennessee. I have attended all of these meetings that my schedule would allow.

The first step THEA took to stop STC was to put out an alert with a resolution in over seven chapter newsletters, numerous support group newsletters, and via e-mail and fax alerts. They explained to Tennessee homeschoolers that our governor, along with businessmen and four or five legislators and the Department of Education, had usurped the authority given by state law to our General Assembly to oversee education. We requested that homeschoolers call their legislators and the governor and ask that STC be stopped, at least until the legislature could consider, debate, and vote on it.

Another significant early step we took to fight STC was to ask Cathy Duffy to speak about it. Mrs. Duffy has researched School-to-Work and Goals 2000, and her book *Government Nannies* gives one of the best summaries of these programs. In mid-April, I was privileged to drive Mrs. Duffy when she spoke in Bristol, Knoxville, Chattanooga, and Nashville as we made a week-long tour of east Tennessee. In Bristol, three legislators—two state representatives and a senator—came to learn more about STW. They expressed their concern about STW/STC and promised to do what they could to help fight it. The message that Mrs. Duffy gave is simple: STW is the federal takeover of education.

In response to that message, more calls came from the people of east Tennessee to their state legislators and the governor. These calls, along with the others that were coming in from people who are members of the coalition phone chains, put the governor and the Department of Education on notice.

In Tennessee, our legislators, who are given the responsibility by our state law to oversee the system of education in Tennessee, did not know about STW until we started informing them. There were less than a half-dozen legislators who were in the loop that knew of STW before the calls began.

It was an interesting situation when the people of Tennessee began to alert the legislators to a program of which they had no knowledge, but which was being initiated by the governor and the commissioner of the Department of Education. The more calls they received and the more they learned about School-to-Work and School-to-Career, the more the state legislators became concerned.

I helped fight STW by talking to legislators. I took my copy of the State Grant Application, which I see as a contract between the state and the federal government, to the legislators' offices.

I have been informing people who are willing to listen about the total system of STW/STC. I have also been involved in getting an amendment attached to our state budget that would stop funding for STC in Tennessee.

Saturday morning, May 31, the legislature recessed until January, 1998. In the last two weeks of the session, the Tennessee Education Information Project had three opportunities to testify before Senate committees concerning STC. These were very important opportunities to inform and thereby impact our legislators on STC.

The coalition flew Mrs. Diana Fessler to Nashville from Ohio to speak to the Senate Education Committee about STW/STC on May 21. Mrs. Fessler is a nationally recognized expert on STW/STC, a member of the Ohio State Board of Education, and a Christian homeschooling mom. (Her web sight is www.fessler.com, if you would like more information including an excellent research paper on STW.) I picked her up at the airport on the 20th of May, drove her downtown to her hotel near the Capitol, and was blessed to spend the next two and half days with her.

The testimony that she gave was so radical, but totally based on fact, that the senators on the committee and the others in the committee room reacted like a grenade went off. It was fascinating to see the damage control that occurred after the testimony. The deputy commissioner of the Department of Education appointed to push STC, the Executive Director of STC in Tennessee, the administration people, and the business partners in STC were all scurrying around like bugs running away from the light. They are the poor, misguided people trying to fix a system based on a worldview that will always fail. However, they think that we're the ones who don't understand and that we are fearful of this program because of our lack of understanding. (This mindset is actually addressed in the grant application; it includes as part of the STC budget an item identified as responding to "fear and misunderstanding.")

Late the same day of Mrs. Fessler's testimony, we had an open meeting in one of the committee rooms to inform interested legislators about STC in Tennessee. Mrs. Fessler spoke against STC along with Mrs. Patray, my father, and Mr. Painter, a local board of education member from one of the most respected school districts in Tennessee. There were twenty legislators in attendance who were very interested in learning about STC. Many not only asked questions but spoke in opposition to STC.

Our next goal was to put an amendment on the appropriations bill (the state budget) to stop the funding from the federal government for STW/STC. In Tennessee, the filing date had already passed for putting an amendment on, but Senator Henry, Chairman of the Finance, Ways and Means Committee, said that he would sponsor the amendment in the Senate. Then he sent us to the House to find a Democrat to sponsor the amendment. By God's grace a representative with whom Mrs. Patray has a long standing relationship signed on as the primary sponsor. I had spoken with him the Thursday before we asked him to sign on and gone over some of our concerns.

On Friday, May 23, Mrs. Patray received a call from Senator Henry's office informing us that the committee would hear testimony and vote on amendments to STC on Monday, May 26, Memorial Day, at 8:30 a.m.

I attended coalition meetings over the weekend to plan our strategy. I was so excited to be a contributing player on the right team. We closed our last meeting with a prayer giving our anxieties over the outcome of the vote the next day to the Lord.

Monday, May 26 was disappointing. The Finance Committee rejected our amendment on a voice vote. However, there were two senators on the committee who hung with us, doing so at a sacrifice to themselves.

On Tuesday, May 27, after I thought all our efforts had failed, Senator Fowler, a Christian homeschooling father from Chattanooga, filed five amendments upon which the Senate would vote on Wednesday.

On Wednesday, May 28 we worked most of the morning before the Senate went to the floor trying to get the seventeen votes we needed to put just one of the amendments on to the budget.

Wednesday afternoon on the floor of the Senate the debate on the first amendment went on for forty minutes. Senator Fowler gave an excellent speech explaining STW/STC. I, Mrs. Patray, and Jonathan Bartley, a homeschooling senior from east Tennessee who also worked at the legislature this year as an intern for THEA, watched as senators lobbied each other on the floor.

When the vote was called for on amendment #5, Mrs. Patray, Jonathan, and I were praying; my stomach was tied in knots. They voted when the bell rang. It seemed like an eternity until the clerk of the Senate called out the vote: 17 ayes, 14 nays, 2 present not voting. We looked in disbelief at one another. God had just performed a miracle. We had just put an amendment on the budget that would stop the second year federal funding for a program that top's the Governor's list of priorities for Tennessee.

However, our victory was quenched on Thursday when a member of the House Republican caucus, Representative Bill Dunn, a homeschooling father from Knoxville, beeped Mrs. Patray while we were in the Senate gallery. He had received an amendment from the administration in a caucus meeting with the governor that would change the wording of amendment #5. Jonathan and I went over to Representative Dunn's office to get a copy of their wording. Representative Dunn said, "Christina, you are one of the smartest people down here. Read this and tell me what you think of their wording." (I realized that the reason he thought of me as being smart was that I have been taught to apply a Christian worldview to all areas of life.)

The budget along with amendment #5 ended up in Conference Committee, a committee appointed by the speakers of each body to work out differences between the House and Senate.

Friday morning, May 30 at 8:00, the conference committee met to publicly do what they had already decided to do in a meeting the night before. Amendment #5, with negotiated wording, was passed on to the budget. The changed wording is better than the original offering by the administration, but it does not stop any funding. It does state that "the General Assembly reserves the right to terminate or continue the acceptance of any federal funds from any grant for a School-to-Career program." The new wording of the amendment also sets up "a joint oversight study committee to investigate and conduct public hearings on STC and report its findings to the General Assembly by February 15, 1998." This means that our legislators will have to investigate and make a decision about the future of STC in Tennessee rather than just let it move ahead without legislative approval.

I have come by my political activism "honestly." My father has been president of THEA for the almost thirteen years since its founding. He was also the first homeschool lobbyist in Tennessee. When I was eight years old, I remember sitting in the Senate committee room playing army men with my brother and another homeschool friend. We were in the Senate Education Committee meeting during the debate of the bill that would include homeschooling as a legal alternative in the State Code. The local news covered the committee meeting and they loved getting a shot of us playing.

Then in 1986, my dad had a creative idea. Bring the homeschoolers to the legislature to meet with their own legislators and show them the quality of homeschool students and the dedication of their parents. So our annual THEA Homeschool Rally/Reception Day began.

Homeschool families come from all across Tennessee to learn about legislation affecting homeschoolers and to meet with their representative and senator. The Rally/Reception Day starts about nine in the morning with speakers throughout the day and time for families to talk with their legislators. In the early afternoon, homeschool teens are given red, white, and blue gift bags full of homemade cookies to give to the staff members. That is my favorite part of the day at the Capitol. The students go in groups of three or four. Somehow, ever since the fifth grade I have been appointed the spokesperson for my group. One year I spoke to the Speaker of the House, Jimmy Naifeh, explaining homeschooling to him. That was one of the highlights of my year.

I am grateful to the Lord and to my parents for their willingness to take the road less traveled. I know that all of the opportunities I am blessed with are not any that I have earned; they have been given to me. I have never attended college. My parents and I have chosen to continue to take that road less traveled. We have been influenced by Mrs. Inge Cannon and by the ATI program that embraces apprenticeship and other nontraditional, historically-proven training.

I know that if I had chosen to go to the local Christian school or if my parents had not chosen to home educate, I would have never had the opportunity to work as an intern for THEA. I would not have assisted this year in the coalition's efforts to stop STC in Tennessee, and I would not have been able to spend time conversing with and learning from godly men and women who are taking a stand to protect our God-given freedoms, men and women such as Cathy Duffy, Diana Fessler, Bobbie Patray, my dad (well, of course, I would have conversed with him!), Senator David Fowler, and Representative Bill Dunn, to name a few. I believe God has given me a lifelong, hands-on involvement in and understanding of the growing grassroots struggle we as homeschoolers have fought in first sealing our initial right to home educate and then moving forward

141

to gain greater freedom from government encroachment. In my opinion, these are definitely activities worth the effort. I am so blessed to be part of the battle for educational freedom. My ultimate goal is to glorify, enjoy, and serve the Lord wherever He calls me.

HOTHOUSE TRANSPLANTS

Brook Tingom, a nineteen-year-old homeschool graduate, lives with her family in Scottsdale, Arizona. Besides tending to her boysenberry patch, practicing the piano and violin, Brook writes for The Home School Digest and publishes Kindred Spirits.

CHAPTER 16

GOD TOOK MY DESIRE AND MADE ME A... WRITER!

When I was a little girl of about seven, I wanted to be a writer when I grew up. I thought it would be delightful to write something others might enjoy! A few years later I still had this view in mind, and I went to a birthday/costume party dressed as a happy mix between the writer and the "written" rolled into one. I donned a pleated plaid skirt and a crisp white blouse (something every writer ought to wear), stuck a pencil behind my ear, and wore two large sheets of paper tied at the shoulders.

Later, I was introduced to grammar and decided I wasn't so interested after all. If writing meant distinguishing nouns, verbs, prepositional phrases and the like, then maybe I wasn't so eager to be a writer. However, God had other plans and later renewed my desire to write, not as an end unto itself, but as a tool for spurring sisters on in the faith.

I am now nineteen years old, and . . . well, I think you've guessed it, I am a writer! God has blessed me with a ministry of publishing and writing a magazine called *Kindred Spirits* for Christian young ladies. It has been over four years now since the first issue was published. My family still marvels at it—God took that desire to write I had even as a little girl—and fulfilled that wish!

I began writing *Kindred Spirits* when I was fifteen years old—at a time in my life when God had given me a deep desire to encourage other girls who love the Lord. I wanted to point to God, encourage young women to live godly lives, and also be friendly and personable. I felt the best way to share this encouragement was to write a small, bimonthly letter to a few friends. That small letter began simply—I did not plan a large magazine or think of it going beyond the small group it went to at the beginning. I just asked God to take it where He willed.

God has abundantly blessed that little ministry and allowed it to grow to a 24-page magazine with a few hundred subscribers. Since the beginning, my vision has been the same, to point young ladies to God in every area of their lives, and to encourage those young women to love and to serve the Lord. So often today, those who desire to live in ways radically different than the world's are looked down upon, but through *Kindred Spirits* I hope to spur those girls on—to not forsake the ways God has directed their families to live. I pray that *Kindred Spirits* will bring together those scattered, like-minded young women for encouragement and edification in the Lord. As a special joy to me, God greatly blesses me from the encouragement I am able to pass on to bless others.

An area *Kindred Spirits* especially upholds is that of biblical womanhood as defined in Titus 2 and illustrated in Proverbs 31—essentially, a young lady remains under her parents' authority until she is given in marriage. In this age of "women's liberation," we desire to uphold God's standard for women as keepers of the home, helpmeets to their husbands, and mothers raising children for the glory of God, as outlined in Scripture. Through *Kindred Spirits* we wish to remind young women that true freedom is following Christ and His design. A woman will be most free and most fulfilled when she denies herself and follows God's design for women. Therefore, the articles and columns are intended to lead young ladies in the ways of learning how to become godly women.

Another closely related scriptural principle we uphold is that God created women to be different than men. God made women for a special purpose in life—for special duties and ways of glorifying Him. These differences are to be rejoiced in, not overcome!

Therefore, we desire to present true femininity and to encourage modesty (versus an attitude of "I'll wear what I want"), a servant's heart (versus a competitive attitude), and a homemaker's spirit (versus a career and independence).

And finally, we hold to the principles of betrothal-courtship rather than dating, and encourage emotional and physical purity. Because I do encourage young women to become homemakers someday, I also wish to provide some practical guidance in learning homemaking skills such as bread-baking, gardening, sewing, and the idea of the hope chest. I hope, in the future, to have some more detailed articles about homemaking in *Kindred Spirits,* such as learning about living on one income, interior decorating, and organization.

Through God's gentle leading my own family came to understand these truths from the Bible and found these standards to be quite unpopular with the rest of the world. Thus, when I felt led to encourage girls, I wanted to speak directly to those young women who were also seeking to live in submission to God in all their ways and who were encountering the same lack of understanding and support we had encountered.

With each issue of *Kindred Spirits* I've learned, little by little, how to write more clearly and how to lay pages out in a more readable format. When I first began, just the thought of writing a one-page article seemed intimidating! There were times when I've stared at a blank piece of paper with lots of thoughts in my head, but found it difficult to convey those thoughts into words.

For me, learning to write did not include long exercises in a grammar book or writing dull essays. I just jumped in and began to write. Every time I write I learn something more about communicating my thoughts so others can understand what I'm thinking. I learn by practice, and I practice because I have a mission to fulfill. Through the help of an older friend who edits all my papers, I've learned how to catch grammatical errors and how to check for flow of ideas and thoughts. This has helped me tremendously.

I've also learned much on the practical side of managing a small ministry/business. Most necessary to the running of this ministry is balancing the finances, keeping up with incoming orders and submissions, and learning how to print and mail the magazine under my budget. I cannot imagine any other way that I could write, edit, publish, advertise, manage a small business (and yet be the mail-opener!), balance finances, consider submissions for publication, and, through all of this, encourage young women to walk in God's ways! I've learned much by the lessons that have sprung up through the years and look forward to all the challenges, as well as the joys and accomplishments that lay ahead. I owe a great deal to my brother, Chris, who is now seventeen years old, for all the help he has given me in working out computer troubles. Chris has a good eye for art and has helped me with page layouts. He also contributes his photography skills for *Kindred Spirits*. Through his experience with working on the magazine, Chris has discovered that he is interested in the field of advertising/marketing and is considering this area for an occupation.

I believe the blessing and opportunity of publishing a magazine came about because of the freedom and preparation I received by being home educated.

In 1983, my mother listened to an audiocassette tape about homeschooling. Her previously skeptical thoughts disappeared, and we became an instant homeschooling family. At the time, I was five years old and had been to a small preschool for two years. After just one month of homeschooling my parents were convinced that it was so beneficial to our family, and had made me so much happier, that they were going to homeschool my brother and me right through high school.

Through the years this resolve was strengthened as they realized that by homeschooling their children they were following God's command given in Deuteronomy that parents teach their children to love and obey God. Homeschooling has been for my parents the necessary means to truly take to heart this God-given responsibility, to carefully guard their children from evil, and to train their children in the way they should go.

Homeschooling has been much more than an academic education. The entire homeschooling lifestyle has played a part in my life that has prepared and aided me in becoming a useful adult much beyond any lesson I could have learned in a textbook. It was not the geography book that gave me an interest in other countries, but listening to the travels of tourists in faraway lands, searching maps in an atlas, and pouring over library books filled with objects strange to my eye that captivated my interest. It was not a history textbook that gave me a love for the past, but the libraries and museums I visited, the actual accounts of our founding fathers that I read, and the stories my grandparents told us of the not-so-far away past that filled my curiosity. It wasn't the textbooks on how to learn that gave me a love for learning. It was the reality of enjoying real learning and investigating life that made homeschooling more than academics from a book. Homeschooling has opened up a world of lessons to learn, an entire way of life that has truly prepared me to step into the work God has given me to do.

Homeschooling also helped make it possible for me to begin *Kindred Spirits* simply because I had enough time to spend on "extracurricular" projects, since I was not tied up with school work all day. My time during high school was structured but not to the extent that it prevented me from being creative or acquiring skills in new interests. I was able to complete my studies and still have time left for serving others and working on projects such as learning to sew, bake, or write stories. And the time available to be creative and inventive has been an invaluable asset to me as each issue of *Kindred Spirits* presents new challenges and changes.

I am so thankful I was homeschooled for all of my school years! I would not have wanted to spend these years any other way. It was just the advantages I mentioned above—the time, creativity, and freedom to learn that homeschooling allows—which made it possible for me to begin *Kindred Spirits* even while I was in high school. Had I not been homeschooled throughout high school, I strongly doubt that I would have begun publishing a magazine.

For many years my family was unsure what I would do after high school. Until about the tenth grade, I thought I would go on to earn a college degree in piano, for I greatly love the piano. However, during my high school years we felt God directing my family not to send me off to college, but instead to continue in the homeschool spirit by learning and serving at home.

Besides publishing *Kindred Spirits*, I enjoy studying other subjects, and I am taking a few college courses through a Christian correspondence program. Though I would like to continue this formal learning, I'm finding that I really only have time for a few courses, as my time is mostly filled with working on the magazine. I also spend my time practicing the piano, teaching a few young piano students, and learning homemaking skills, of which gardening is my favorite (and somewhat of a challenge here in the hot, dry southwest!).

In the last year, my mother and I have become interested in home redecorating (using what people already own) and have redecorated the homes of several friends. God has given me a lot of interests! At times it has been difficult to know which one He desires to be my main focus, but, as my family and I are just starting to realize, God is leading me to make *Kindred Spirits* my primary focus as it grows and requires more and more time. I am so thankful for this ministry and am delighted to see His guiding hand direct my life.

Looking back I can see how God was preparing me for this ministry. The first part of the preparation was giving me something worth saying. God has been building "something worth saying" in me throughout my life—by placing me in the family in which I could grow, but more particularly, by grounding me in His Word through much Bible memorization. I was willing to spend great quantities of time daily memorizing Scripture, and God used that to teach me about Himself and His ways. Through this Bible memorization He was also building in me a desire to follow Him, and not the world's ways.

The second part of preparation was God giving me the wish to encourage other girls to love God with all their hearts and a desire to be a friend to those leading lives separate from the world for Him.

And now you are probably expecting me to say, "The third part of preparation was that God gave me a talent for writing"—am I right? But actually, I really had no talent for writing. I merely had something I felt I needed to write. I had not even properly studied English grammar—we had just never fit it into our studies. But I learned quite a bit soon after beginning *Kindred Spirits* with my mother's instructions. My mother achieved just the right balance of allowing me to write despite my mistakes, while adding certain lessons in grammar that I did need to learn. If she had been overly zealous in making me present perfect papers or spend much time in a grammar book, I might not have developed a heart for writing. Yet if she had let me get by with less than I could have done, I would not have come as far. (By the way, I am now "properly" studying English grammar through correspondence college.)

And here I am four years later. God has been good to me and blessed *Kindred Spirits* much beyond what I could have expected. He has been faithful to keep my mailbox filled with all the material I need for each issue, and has led it to the girls whom it would bless. In my personal life,

God has been so gracious to me! He continues, to build my character with each issue, leads me to trust Him more, and stretches me in my writing. I feel I am a vessel of encouragement to those young women who also love Him, and I have seen that the best encouragement comes when I empty myself of all my desires, all my wants, and all of my self, and just let God work through me. After all, I am nothing, and couldn't hope to bless even one person by myself—but I can hope to be a blessing to many, if it is truly God working through me.

Kindred Spirits is a bimonthly magazine written to encourage Christian young ladies, approximately ages 10-25, to "walk in Jesus' steps." Sample issues are available for $2, and subscriptions for $12. Write to *Kindred Spirits* Dept. 6, 6628 E. Beryl Ave., Scottsdale, AZ 85253-1330.